Battlefield Forensi
Persian Gulf States
Regional and U.S. Military Weapons, Ammunition and Headstamp Markings

Battlefield Forensics for Persian Gulf States

Regional and U.S. Military Weapons, Ammunition and Headstamp Markings

By
Don Mikko and William Bailey

CRC Press
Taylor & Francis Group
Boca Raton London New York

CRC Press is an imprint of the
Taylor & Francis Group, an **informa** business

CRC Press
Taylor & Francis Group
6000 Broken Sound Parkway NW, Suite 300
Boca Raton, FL 33487-2742

Library of Congress Cataloging-in-Publication Data

Names: Mikko, Don, author. | Bailey, William S. (William Scherer), author.
Title: Battlefield forensics for Persian Gulf States : regional and U.S.
military weapons, ammunition and headstamp markings/authored by Don
Mikko and William Bailey.
Other titles: Regional and U.S. military weapons, ammunition and headstamp markings
Description: Boca Raton, FL : CRC Press/Taylor & Francis Group, [2019]
Identifiers: LCCN 2019000327| ISBN 9781138370593 (hbk. : alk. paper) | ISBN
9781138370609 (pbk. : alk. paper)
Subjects: LCSH: Forensic ballistics--Persian Gulf States. | Forensic
ballistics--United States. | Middle East--Armed
Forces--Firearms--Markings. | United States--Armed
Forces--Firearms--Markings. | Firearms--Identification. |
Firearms--Persian Gulf States.
Classification: LCC HV8077 .M46 2019 | DDC 623.4--dc23
LC record available at https://lccn.loc.gov/2019000327

Visit the Taylor & Francis Web site at
http://www.taylorandfrancis.com

and the CRC Press Web site at
http://www.crcpress.com

Contents

Foreword

There are countless books that discuss military firearms, from encyclopedia-sized books that cover every possible gun used by each army throughout history to books that are devoted to one specific firearm. The intent of Don Mikko and William Bailey, the coauthors of this particular book, was to focus on the firearms that are most likely to be encountered on any modern battlefield or in any contemporary military maneuver. Their idea was to put this material into a compact, convenient format so it is informative, is enjoyable to browse, and can be utilized by anyone, from the battlefield investigator to a military gun enthusiast.

Being an expert among experts within the field of military firearms and forensics for more than 35 years, Don Mikko puts a wealth of firearms information that is both technical and practical into the readers' hands.

What sets Don apart from other experts in the battlefield firearms forensics world is the combination of his experience, his specialized training, and his many years of involvement in unique assignments within the U.S. military and the forensic firearms community. From his position as a special agent with the U.S. Army Criminal Investigation Command (USACIDC) (CID) to his role as the branch chief of the Firearms and Toolmark Section at the U.S. Army Criminal Investigation Laboratory (USACIL), Don has been involved at every level with the weaponry that finds its way to conflicts, both regional and worldwide, along with crime scenes throughout the world.

While serving our military services in the capacity of a forensic firearm and toolmark examiner in Afghanistan, I routinely depended on Don as a resource for detailed technical information on the more unique guns and ammunition collected from the battlefield. Not only have I been fortunate to draw upon Don's extensive knowledge for many years, but also I consider Don to be a close personal friend and an esteemed colleague with the highest standards of honesty, integrity, and personal self-sacrifice. Don recognized the need to have a compact resource book that can be used in every environment, from classroom training to the deserts and jungles where the very guns covered in this book are being used in insurgencies and military actions today. I have no doubt that many an investigator, whether wearing a business suit, a lab coat, or a military uniform, will find this book to be an essential asset in their investigations and research. I am sure this pocket-sized book will be carried in everything from police vehicles, jeeps, and armored vehicles to military aircraft and warships at sea. In summary, this indispensable and valuable reference guide is the latest of Don Mikko's many contributions to battlefield forensic firearms investigations.

Edward "Kevin" Lattyak

Edward "Kevin" Lattyak is currently the supervisor of the firearms section at the Cuyahoga County Office of the Medical Examiner, Crime Laboratory, Cleveland, Ohio. Kevin has more than 36 years of experience as a forensic firearms and toolmark examiner who has worked in various crime laboratories within the United States and as a contractor within the combat theater of Afghanistan. Like Don, Kevin is a court-certified expert within his profession and has served his country honorably and faithfully.

Preface

Forensics is defined as the study and practice of the application of science to the purpose of the law (Forensic Science Foundation). The battlefield's primary uses for forensics include the examination of crime scenes and the gathering of forensic evidence to be subsequently examined and potentially used for either criminal or civil prosecution within combat operations.

Throughout my military and federal government service career, I have investigated and/or examined firearms-related items of evidence in countless crime scenes throughout the world. Additionally, during my tenure as the chief of the Firearms Branch at the U.S. Army Criminal Investigation Laboratory (USACIL), Fort Gillem, Georgia, I was also responsible for the recruitment, training, and competency testing of numerous military contractors deploying to various combat-related areas of the world, providing forensic firearm and toolmark identification support to the local commanders.

Due to the fact that CID agents, contractors, crime scene technicians, and military personnel were being deployed to the combat theater where various foreign weapons and ammunition were being utilized by the enemy, I felt it was imperative to educate them on the technical data of different types of firearms, ammunition, county of origin, proofmarks, penetrator information, and typical firearm markings imparted on fired cartridge cases, such as breech face, firing pin, ejector, and extractor markings.

Forensics and biometrics within the Persian Gulf have been leveraged since the early 1990s, and many of the firearms and types of ammunition covered in this handbook are addressed in an effort to assist the professionals on the ground. The information contained within the handbook has already proven to be effective on the battlefield and has helped shape combat operations. The scope of the battlefield in Iraq alone has expanded beyond improvised explosives and weapons. As a result, general support laboratories were established beginning in 2005 in various locations within Iraq, Afghanistan, and Africa.

This handbook will not only assist military and contractor personnel within the various combat locations, but also prove to be beneficial for law enforcement, crime laboratory personnel, forensic firearm and toolmark examiners, crime scene investigators, students, practitioners, and civilian personnel throughout the world. Many of the firearms addressed in this handbook have been either illegally smuggled back to the United States or redesigned and sold in various platforms, including the semiautomatic mode of fire.

Debt of Gratitude

Special thanks to Ms. Ayanna B. Echols, EdS, Mr. Katanga Bailey, Mr. Ryan Coffey, Ms. Dana Bonar Gicale, Mr. Jerry J. Miller, Mr. Ryan Morrison, Mr. Gordon Overholtzer, Mr. Richard "Van" Roberts and a debt of gratitude to the instructors and staff of the former Opposing Forces (OPFOR) and Explosive Ordnance Disposal (EOD) Units at Aberdeen Proving Grounds, MD, Baumholder Germany, Fort Bragg, NC, Fort Campbell, KY, Fort Gillem, GA, Fort Knox, KY, Fort. McClellan, AL, and Fort Riley, KS.

Special thanks to the men and women of our Armed Forces whom I've proudly served with throughout my years of active duty service along with thousands of forensic experts and law enforcement personnel, both stated and unknown. You have inspired me to be the best at what I do within my profession.

Don Mikko

Personal thanks to Don Mikko for being my teacher/mentor in the vocation of Forensic Firearms and Toolmark Identification and providing me with the necessary skills sets in this specialized field of forensic science. I would also like to thank my wife for allowing me to pursue my hobby which has now become my vocation.

William Bailey

About the Authors

Don Mikko is currently the president of Forensic Firearms Training Seminars, Inc., which provides a variety of forensic services for attorneys, law enforcement agencies, insurance companies, private investigators, military organizations, international organizations, college students, and civilian personnel. For more information, visit www.mikkoforensics.com.

He's also a retired chief warrant officer who served in the U.S. Army and spent most of his military career as a special agent in the U.S. Army Criminal Investigation Command (USACIDC).

Don has been a forensic firearm and toolmark examiner since 1990 and has spent 22 years at the U.S. Army Criminal Investigation Laboratory (USACIL), Fort Gillem, Georgia, where he was the chief of the Firearms Branch. He has also served as the first laboratory director at the Atlanta Police Department Forensics Crime Laboratory, Atlanta, Georgia.

He's an American Society of Crime Laboratory Directors/Laboratory Accreditation Board (ASCLD/LAB) inspector and ISO 17025 assessor and is certified by his professional organization, known as the Association of Firearm and Toolmark Examiners (AFTE), in all three forensic disciplines: firearms, toolmarks, and gunshot residues. He has served on several AFTE committees and is currently a member of the AFTE Certification Committee.

He's also an adjunct professor at Southern Crescent Technical College, Griffin, Georgia, and has instructed forensic and criminal justice courses at Central Texas College, Baumholder, Germany; Clayton State College, Morrow, Georgia; Sirchie, Youngsville, North Carolina; and Central Piedmont Community College, Charlotte, North Carolina.

Don is a graduate of the FBI National Academy Association (FBINAA), 180th Session (1995). He has a master's degree in business management from the University of Phoenix, a bachelor's degree in criminal justice from Chaminade University, and an associate's degree in police science from the University of Hawaii.

Don is post certified within the state of Georgia for several forensic firearm type courses and is also a member of several professional organizations, including the AFTE, Internal Association for Identification (IAI), International Ammunition Association (IAA), CID Agents Association (CIDAA), FBINAA, 50 Caliber Shooters Association, National Rifle Association (NRA), and the Georgia Law Enforcement Firearms Instructor Association (GALEFI).

As a special agent, he has testified in various military and civilian courts in excess of 200 cases throughout his professional career.

Today, Don spends most of his time reexamining evidence in various criminal and civil type cases and is very engaged in various training assignments for crime laboratories, police departments, military organizations, and college students.

William Bailey is president of Bailey Firearm Forensics, Inc., which is a forensic training and consulting corporation responsible for providing forensic support to attorneys, law enforcement agencies, insurance companies, private investigators, military organizations, and civilian personnel. He is also a forensic firearm and toolmark examiner. Some of his duties include the examination of firearms; the examination of ammunition and ammunition components; the microscopic comparison of questioned bullets, cartridge cases, and toolmarks; the restoration of obliterated serial numbers; distance determinations (both gunpowder and shotgun pattern analysis); the examination of security devices such as padlocks and safes; and shooting reconstruction. Additionally, William has a degree in Electrical Engineering (BSEE) and has developed software for various large and small companies for over 20 years.

Assistant Editors

Raymond Barnwell
Firearms Supervisor, York County Sheriff's Office, York County, SC

Albert D. Bell
Forensic Firearm and Toolmark Examiner

Aaron Fullerton
Forensic Firearm and Toolmark Examiner

Donna J. Jackson
Supervisor, Durham Forensics Laboratory, Durham, NC

George Kass
Forensic Ammunition Service, Inc., Okemos, MI

Ed Love
Forensic Firearm and Toolmark Examiner

Tammy Starckey
Forensic Firearms Training Seminars, Inc.

Dean Wriston
Forensic Firearm and Toolmark Examiner

Special thanks to all the forensic firearm and toolmark examiners and contractors for their assistance throughout the years.

Weapons, all gold plated, taken from Saddam Hussein's palace: Top –
Al Kadesih sniper rifle; bottom left – AKMS-U; bottom right – Tariq
pistol.

Introduction

This book is designed to facilitate the identification, functionality, and country of origin of the various firearms/weapons encountered in the Middle East, Persian Gulf, and other foreign countries.

Although much of the equipment listed has been produced as early as the 1950s, and in some cases, such as the AK-47, the late 1940s, their existence was not quite as relevant as after September 11, 2001, when the World Trade Center and Pentagon were attacked. Although the United States had been involved in proxy wars between various countries such as Iraq, Iran, and Israel, this involvement became direct and personal when the author of the 9/11 attack was found to be residing in, and abetted by, Afghanistan.

The resultant invasion then pitted the U.S. military and allies directly against those either living in or politically or religiously associated with Osama Bin Laden.

Many of the weapons encountered in the Middle East and Persian Gulf, though not all, were and are supplied by the former Soviet Socialist Republic of Russia and People's Republic of China. Due to the proven simplistic, dependable functionality, combined with inexpensive and readily available ammunition, these firearms are ubiquitous within the region. As a result of the availability of certain firearms trumping best choice, many other types and origins of firearms may be encountered. These firearms, though less common, are always found to be using ammunition that is easy and cheap to supply, specifically 5.56 × 45mm (NATO), 9 × 19mm (NATO), 7.62 × 39mm, 7.62 × 51mm (NATO), and 7.62×54mmR.

Regionally produced firearms and ammunition of the same caliber are also likely to be encountered. Additional arms are also encountered being used against targets ranging from automobiles to aircraft. These are predominantly Soviet in origin, such as rocket-propelled grenades (RPGs) and machine gun-type grenade launchers, as well as large-caliber heavy machine guns.

The invasion of Iraq led to the destabilization of the region and a civil war between the two dominant factions in Islam: Sunni and Shia. This precipitated a civil war that has resulted in vast amounts of civilian casualties. Another result is that much of the ammunition and money being supplied by nations are aligned with one of these two religions. Subsequently, many of the firearms and munitions will bear the lettering/labeling of these nations.

One aim of this publication is to aid firearms examiners in the field in determining what weapon is used, based on caliber, ammunition type and caliber, origin of ammunition, and whole or parts of the projectile recovered.

Another point of interest is to be able to operate a weapon that the examiner or user has not previously encountered in a safe manner to evaluate possible involvement in a firefight or crime, or indeed, if the firearm is even operational.

This book can be used, then, to determine how various firearms are operated and what type of ammunition they use. This book can also be used to determine the origin of the ammunition producing resultant metal cores or penetrators, based on shape, caliber, weight, and length.

This book can be used to determine the origin of manufacture of unfired and fired ammunition based on the headstamp found on the base of the cartridge/cartridge case.

It can be used by law enforcement agencies in the event that any of the firearms listed are encountered, either in the perpetration of domestic crime or in the activity of domestic terrorism.

In closing, this book can also be used by the curious, collectors, and interested individuals who wish to remain well informed, as well as lending assistance in properly identifying the make, model, and manufacture of these specialized, uncommon firearms. Along with proper identification of foreign firearms, the book also can assist with the identification of various types of foreign ammunition.

Weapons of the Opposing Forces in the Persian Gulf, Middle East, and Other Foreign Countries

I

The weapons encountered in the Middle East, Persian Gulf, and other foreign countries are far ranging in type, caliber, origin, and application. These devices include handguns, selective fire rifles, sniper and assault rifles, submachine guns, and heavy machine guns, as well as various delivery systems for grenades used against targets as diverse as the foot soldier, armored personnel carriers, tanks, and aircraft.

This section outlines those weapons commonly encountered. The following information includes detailed images for definitive identification, ammunition types and applications, rate of fire, barrel's rate of twist, effective range, origin of manufacture, and other operational characteristics, as well as physical characteristics such as weight and length of the barrel/overall weapon. Also included is a listing of strengths and weaknesses for each, a brief summary of functionality, and precautionary information.

Contents

I.1   Makarov 9mm Pistol (PM)

TECHNICAL DATA

Caliber: 9 × 18mm Makarov
Operation: Double-action, Blowback, semiautomatic
Max eff range: 40 m
Feed: 8 rd box mag
Rate of fire: 40 rds/min
Weight: 1.7 lbs.
Length, overall: 6.34 in.
Length, barrel: 3.83 in.
Rifling: 4 right
Rate of twist: 1:9.5 in.
Land width: 0.084–0.092 in.
Groove width: 0.180–0.184 in.
Extractor: 2 o'clock
Ejector: 7 o'clock
Manufacturer: Izhevsk Mechanical Plant, Soviet Union

Weapons Brief

The Makarov 9mm pistol (PM) was produced in the former Soviet Union and was the standard firearm within their exhaustive inventory. The Makarov can easily be recognized by the lanyard loop and encircled star located on the grips. The serial number is located on the left side of the frame and on the left side of the slide. The safety is located on the left side of the slide, just in front of the hammer. The Makarov has a double-action trigger mechanism and is semiautomatic.

Also, the Makarov typically uses a low-powered cartridge that subsequently reduces its overall range and penetration power.

The former East German government called this weapon the Pistole M and the People's Republic of China (PRC) calls it the Type 59 Shi. The Pistole Model has regular handgrips without a landyard loop while the PRC version has "59 SHI" stamped on the receiver.

Many of these weapons were seized by various personnel on the battlefield and subsequently smuggled to the United States.

Operability

1. Visually check that the chamber is clear.
2. Visually check the barrel to ensure that there is no obstruction.
3. Insert the magazine.
4. Place the safety lever in the fire position.
5. Pull the slide to the rear and release.
6. The Makarov is now ready to fire.

I.2   Tariq 9mm Pistol

TECHNICAL DATA

Caliber: 9×19mm, .32 ACP (7.65×17mm)
Operation: Single-action, semiautomatic, short recoil
Max eff range: 50 m
Feed: 9mm 8/11 rd box mag,
    .32 ACP 8 rd box mag
Rate of fire: 40 rds/min
Weight: 2.15 lbs.
Length, overall: 8 in.
Length, barrel: 4.5 in.
Rifling: 6 right
Rate of twist: 1:10 in.
Land width: 0.044–0.048 in.
Groove width: 0.140–0.146 in.
Extractor: 2–3 o'clock
Ejector: 9 o'clock
Manufacturer: Al Qadisiyah Governorate, Iraq

Weapons Brief

The manufacture of the Tariq pistol began during the late 1970s after Beretta sold rights and tooling for the Model 1951 to Iraq. The pistol subsequently underwent only minor cosmetic changes, such as the Iraqi logo on the hand grips and the Iraqi writing on the right side of the slide. The left side of the slide has the printing "TARIQ 9m/m IRAQ-Licensed by Beretta." The models issued to Saddam Hussein's private police have a special marking on the frame.

The safety located at the upper rear of the hand grips is a trigger block mechanism, which can be placed on safe only when the firearm is cocked. There is also a half-cock safety. The serial number is located on the left side of the frame, forward of the trigger guard beneath the slide. Production of Tariq pistols has been limited. Additionally, replacement/spare parts are very difficult to locate.

The Tariq 9mm pistol has been in service with the Iraqi Police, the Iraqi Armed Forces and the Republican Guard.

Many of these weapons were seized by various personnel on the battlefield and subsequently smuggled to the United States.

Operability

1. Visually check that the chamber is clear.
2. Visually check the barrel to ensure that there is no obstruction.
3. Insert magazine.
4. Push safety to the fire position.
5. Pull the slide rearward and release.
6. The Tariq is now ready to fire.

I.3   Hungarian PA-63 Pistol

TECHNICAL DATA

Caliber: 9 × 18mm Makarov, .32 ACP, .380 ACP
Operation: DA, semiautomatic, blowback
Max eff range: 50 m
Feed: 7 rd box magazine
Rate of fire: 35 rds/min
Weight: 1.5 lbs.
Length, overall: 6.75 in.
Length, barrel: 4 in.
Rifling: 6 right
Rate of twist: 1:9.45 in.
Land width: 0.072–0.078 in.
Groove width: 0.108–0.116 in.
Extractor: 3 o'clock
Ejector: 9 o'clock
Manufacturer: FEGARMY Arms Factory Ltd, Budapest, Hungary

Weapons Brief

During Operation Desert Storm, variants of the Hungarian PA-63 were found in use by the Iraqis. Unlike the Model 74, the model PA-63s found in the desert were manufactured in Hungary then exported to Iraq. The PA-63 can be identified by "PA-63" stamped in English on either side of the slide or frame. The serial number containing two letters followed by four numbers is on the slide, frame, and magazine. The Hungarian provisional and voluntary proof stamp (a capital I within an antler-like design) appears on the frame. The safety is a rotary-type de-cocking lever that is located on the left side of the frame and in front of the hammer. The safety/de-cocking lever, which is located on the rear of the slide, must be pushed forward to disengage the safety. Due to its overwhelming popularity and durability, FEG later manufactured models in .32 ACP and .380ACP.

Many of these weapons were seized by various personnel while on the battlefield and subsequently smuggled to the United States.

Operability

1. Visually check that the chamber is clear.
2. Visually check the barrel to ensure that there is no obstruction.
3. Insert magazine.
4. Rotate safety lever to the fire position.
5. Pull the slide rearward and release.
6. The pistol is now ready to fire.

I.4   Iraqi Model 74 Pistol

TECHNICAL DATA

Caliber: .32 ACP (7.65 × 17mm)
Operation: DA, semiautomatic, blowback
Max eff range: 50 m
Feed: 8 rd box magazine
Rate of fire: 40 rds/min
Weight: 1.29 lbs.
Length, overall: 6.25 in.
Length, barrel: 3.5 in.
Rifling: 4 right
Rate of twist: 1:10 in.
Land width: 0.082–0.088 in.
Groove width: 0.148–0.156 in.
Extractor: 3 o'clock
Ejector: 9 o'clock
Manufacturer: FEGARMY Arms Factory Ltd, Budapest, Hungary

Weapons Brief

The model 74 is a copy of the Walther PP. The parts for the Model 74 were delivered to the Iraqis unassembled. The Model 74s found on the battlefield by various troops in the Middle East were usually of poor quality.

The Model 74 is manufactured in the calibers of .32 ACP (7.65 × 17mm), .380, and 9 × 18mm Makarov. The serial number is located toward the left side of the slide and frame to the rear of the trigger. The safety lever is a rotary type and is located at the left side of the slide in the rear of the slide grip. The action of Iraq Model 74s is very rough and rigid. The Model 74s found by Allied Forces in the Middle East were generally of poor quality.

During the First and Second Gulf Wars numerous Model 74 pistols were seized by various personnel while on the battlefield and subsequently smuggled to the United States. A majority of these firearms were in poor and non-operational condition.

Operability

1. Visually check that the chamber is clear.
2. Visually check the barrel to ensure that there is no obstruction.
3. Insert magazine.
4. Rotate safety lever to the fire position.
5. Pull the slide rearward and release.
6. The pistol is now ready to fire.

I.5 Soviet AK-74 Assault Rifle

courtesy wiki commons

TECHNICAL DATA

Caliber: 5.45 × 39mm
Operation: Gas, selective fire
Locking system: Rotating bolt
Max eff range: 400–500 m (sight adjustment)
Feed: 30 rd box mag, 45 rd box, 75 rd casket drum
Rate of fire: 650 rds/min
Weight: 6.7 lbs.
Length, overall: 37.1 in.
Length, barrel: 16.3 in.
Rifling: 4 right
Rate of twist: 1:8 in.
Land width: 0.074–0.078 in.
Groove width: 0.092–0.096 in.
Extractor: 2 o'clock
Ejector: 9 o'clock
Manufacturer: Izhevsk Mechanical Plant, Soviet Union

NOTE: Some specifications may vary according to country of origin.

Weapons Brief

The AK-74 fires a 5.45 × 39mm cartridge. It's easily recognizable by its large muzzle compensator on the barrel, which is designed to direct recoil straight back into the stock while thus reducing the muzzle creep effect when firing in the fully automatic mode of fire. The magazine is made of an orange-colored fibrous plastic (Bakelite) that has a steel sleeve. It also has a wood stock and a plastic pistol type grip.

The serial number is located on the left side of the receiver and adjacent to the chamber. The safety lever is located on the right side of the receiver and is similar to the AK-47/AKM.

1. The AK-74 is a very reliable assault rifle which was manufactured along the line of the AK-47/AKM design.
2. The AKS-74U version has a barrel that is slightly shorter and is equipped with various types of folding stocks.
3. The AKS version also has a folding metal stock and many of them were also found on the battlefield during recent conflicts and subsequently smuggled to the United States and other countries.

Operability

1. Visually check that the chamber is clear.
2. Visually check the barrel to ensure that there is no obstruction.
3. Ensure that the weapon is on safe and cleared.
4. Place selector lever in the semiautomatic or fully automatic mode of fire.
5. Insert a loaded magazine.
6. Pull the charging handle rearward and release.
7. Pull trigger when ready to fire.

I.6 Soviet AK-47/AKM Assault Rifles

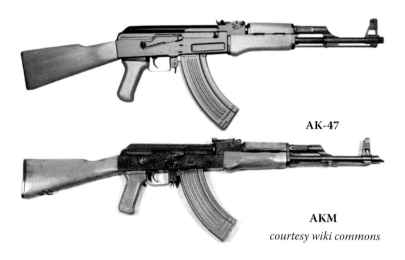

AK-47

AKM
courtesy wiki commons

TECHNICAL DATA

Caliber: 7.62×39mm, M43
Operation: Gas, selective fire, semiautomatic and automatic
Max eff range: 300–400 m (sight adjustment)
Feed: 30, 40 rd box/75 rd RPK drum mag
Rate of fire: 600 rds/min
Weight: (AK-47) 9.5 lbs.
          (AKM) 6.83 lbs.
Length, overall: 34.3 in. (both)
Length, barrel: 16.3 in. (both)
Rifling: 4 right
Rate of twist: 1:9.25 in.
Land width: 0.085–0.088 in.
Groove width: 0.130–0.134 in.
Extractor: 3 o'clock
Ejector: 9 o'clock

Manufacturer: Numerous countries of origin (Izhevsk Mechanical Factory)
NOTE: Some specifications may vary according to country of origin.

Weapons Brief

The AK-47/AKM assault rifles have long been the primary weapons of the Soviet and former Warsaw Pact countries. The primary differences between the AK-47 and the AKM are weight and construction. The metal parts of the older AK-47 are machined and the AKM parts are stamped. All Kalashnikov assault rifles in this series fire the Soviet 7.62×39mm M1943 cartridge. The serial number is located on the left side of the receiver and adjacent to the chamber. The manual safety is a selector lever located on the right side of the receiver. In the upward position the selector is on safe, the middle position is automatic mode of fire, and the downward position semiautomatic mode of fire.

Many of these weapons were seized by various personnel on the battlefield and subsequently smuggled to the United States.

Operability

Operation of the AK-47/AKM is as follows:

1. Visually check that the chamber is clear.
2. Visually check the barrel to ensure that there is no obstruction.
3. Insert a loaded magazine.
4. Place the selector, located on the right side of the weapon, on semiautomatic (down) or automatic (up).
5. Pull the charging handle rearward and release.
6. Pull trigger when ready to fire.

I.7   UZI Submachine Gun

TECHNICAL DATA

Caliber: 9 × 19mm, .22LR, .45ACP, .41AE, 9 × 21IMI
Operation: Blowback, selective fire, open bolt
Max eff range: 200 m
Feed: 25, 32, or 40 rd box mag
Rate of fire: Semi, 40/50 rds/min
            Effective, 80/100 rds/min
            Cyclic 600 rds/min
Weight: 7.72 lbs.
Length, overall: 25.6 in.
Length, barrel: 18.5 in.
Rifling: 4 right
Rate of twist: 1:10 in.
Land width: 0.118–0.122 in.
Groove width: 0.151–0.154 in.
Extractor: 2 o'clock
Ejector: 7 o'clock
Manufacturer: IMI Systems (formerly Israeli Military Industries)

Weapons Brief

The 9mm UZI submachine gun was designed by Uziel Gal, an Israeli army officer, in the early 1950s after the Arab–Israeli War. The UZI has two unique features: a bayonet catch and a grenade launcher attachment. It can be fitted with wooden or metal folding stocks.

The serial number is located on the right side of the receiver. It was engineered with safety in mind and designed with a grip safety to prevent accidental firing if the weapon is dropped, as well as designed with no external moving parts to cause injury when being fired.

The UZI is also sold commercially in many countries throughout the world and is still used by many terrorist organizations outside the United States. Numerous UZI weapons were seized by various personnel on the battlefield and smuggled to the United States.

Operability

1. Visually check that the chamber is clear.
2. Visually check the barrel to ensure that there is no obstruction.
3. To fire, place the selector in the forward (automatic) or middle (semi-automatic) position.
4. Press and hold the grip safety (similar to the M1911A1 .45 caliber pistol models) and pull the operating handle fully to the rear, and then release it.

CAUTION: To fire the UZI, depress the grip safety and squeeze the trigger. If the weapon is not to be immediately fired, place the selector fully forward to the safe position.

I.8   UK L2A3/L34A1 Submachine Guns

TECHNICAL DATA

	L2A3	L34A1
Caliber	9 × 19mm	9 × 19mm
Operation	Blowback, selective fire, open bolt	
Max eff range	200 meters	50–100 meters
Rate of fire	550 rds/min	550 rds/min
Weight, loaded	8 lbs.	10 lbs.
Length, folded	19 in.	26 in.
Length, extended	27 in.	34 in.
Rifling	6 right	6 right
Rate of twist	1:10 in.	1:10 in.
Land width	0.058–0.061 in.	0.075–0.080 in.
Groove width	0.120–0.123 in.	0.098–0.101 in.
Extractor	3 o'clock	3 o'clock
Ejector	9 o'clock	7 o'clock
Manufacturer	Sterling Armament Company, Dagenham, UK	

Weapons Brief

The serial numbers for the L2A3 and L34A1 are located on the top of the magazine well. The safety lever on both weapons is a rotary type that is located on the left side of the receiver just behind the trigger.

Firing the L34A1 in the automatic mode can force excessive gas into the suppressor which may cause permanent damage. The L34A1 has an unperforated barrel that holds the gases in and causes the weapon to overheat.

Both weapons can be found in the Canada, Ghana, India, Malaysia, New Zealand, United Kingdom, and several other countries. Numerous versions of these weapons were seized by various personnel on the battlefield and smuggled to the United States and other countries.

Operability

1. Visually check that the chamber is clear.
2. Visually check the barrel to ensure that there is no obstruction.
3. Put the safety lever in the safe position.
4. Insert the magazine.
5. Pull the charging handle to the rear and release.
6. Place the selector lever in the semiautomatic or automatic mode of fire and begin firing.

I.9   German G3 Rifle

TECHNICAL DATA

Caliber: 7.62 × 51mm NATO
Operation: Roller-delayed blowback, selective fire
Max eff range: 400 m (sight adjustment)
Feed: 20 rd box magazine
Rate of fire: Semi 60 rds/min
                    Effective 100 rds/min
                    Cyclic 500–600 rds/min
Weight: 9.7 lbs.
Length, fixed stock: 40 in.
Length, retractable: 33 in. (closed)
Length, barrel: 18 in.
Rifling: 4 right
Rate of twist: 1:12 in.
Land width: 0.054–0.057 in.
Groove width: 0.148–0.152 in.
Extractor: 2 o'clock
Ejector: 6 o'clock
Manufacturer: Heckler & Koch, Germany

Weapons Brief

The German G3 (7.62 × 51mm) assault rifle was designed from the Spanish Centro de Estudios Tecnicos de Materiales Especiales (CETME) rifle and had been adopted as the standard rifle by the former West German Army. The serial number is located on the left side of the lower receiver, near the magazine well. The safety lever is a rotary type, located on the left side of the receiver, just above the trigger.

The standard rifle known as the G3A3 has four variants: the G3K, G3A4, G3A3ZF, and G3SG/1.

The armed forces and/or police forces of Africa, Central and South America, the Middle and Far East, and of course throughout Europe have used the G3 rifle.

Operability

1. Visually check that the chamber is clear.
2. Visually check the barrel to ensure that there is no obstruction.
3. Insert the magazine.
4. Rotate the selector upward to the safe position.
5. Pull the operating handle to the rear and release it.

I.10   Soviet PPS-43 Submachine Gun

TECHNICAL DATA

Caliber: 7.62 × 25mm, Tokarev M1930
Operation: Blowback, open bolt
Max eff range: 200 m (sight adjustment)
Feed: 35 rd box mag
Rate of fire: Auto 650 rds/min
           Effective 100 rds/min
           Cyclic 600–700 rds/min
Weight: 6.7 lbs.
Length, extended: 32.3 in.
Length, folded: 24.2 in.
Length, barrel: 9.6 in.
Rifling: 4 right
Rate of twist: 1:9.6 in.
Land width: 0.080–0.083 in.
Groove width: 0.150–0.154 in.
Extractor: 3 o'clock
Ejector: 9 o'clock
Manufacturer: Izhevsk Mechanical Plant, Soviet Union

Weapons Brief

Weapon that fires in the automatic mode of fire only. The PPS-43 has a barrel jacket, which extends beyond the barrel, which is designed to prevent direct contact.

The receiver is also stamped metal and is equipped with a folding stock. The PPS-43 won't accept a drum magazine. The serial number is located on top of the receiver, directly behind the ejection port. It also has an external lever safety that prevents accidental discharge. When placed in the safe position, both the bolt and trigger are disabled. The safety is located on the right side of the receiver, just in front of the trigger guard.

Numerous versions of these weapons were seized by various personnel and subsequently smuggled to the United States.

Operability

1. Visually check that the chamber is clear.
2. Visually check the barrel to ensure that there is no obstruction.
3. Insert magazine.
4. Pull charging handle to the rear and release.
5. The weapon is now ready to fire.

I.11   Soviet PPSh-41 Submachine Gun

TECHNICAL DATA

Caliber: 7.62 × 25mm Tokarev
Operation: Blowback, open bolt
Max eff range: 125–150 m
Feed: 35 rd box mag, 71 rd drum
Rate of fire: Semi 40 rds/min
        Effective 105 rds/min
        Cyclic 900 rds/min
Weight: 8 lbs.
Length, overall: 33 in.
Length, barrel: 11 in.
Rifling: 4 right
Rate of twist: 1:9.5 in.
Land width: 0.079–0.081 in.
Groove width: 0.160–0.164 in.
Extractor: 3 o'clock
Ejector: 9 o'clock
Manufacturer: Izhevsk Mechanical Plant, Soviet Union

Weapons Brief

The Soviet PPSh-41 stands for Pistolet Pulemyot Shpagina first manufactured in 1941. The weapon has a wooden butt stock; the barrel is chromium plated. The barrel jacket extends beyond the barrel and is sloped back from top to bottom and vented on all sides. The barrel jacket acts as a muzzle brake and compensator.

The serial number is located on top of the receiver, just behind the ejection port. It is also located on the bottom of the trigger guard. The safety, which is a slide release lever, is located on the bolt carrier charging handle.

The PPSh-41 has also been manufactured in the People's Republic of China (PRC) (Type 50), Hungary (48M), Yugoslavia (Type (49), and Vietnam (K50).

Both Iran and former East Germany countries re-chambered the PPSh-41 in order to to fire a 9×19mm cartridge.

Operability

1. Visually check that the chamber is clear.
2. Visually check the barrel to ensure that there is no obstruction.
3. Make sure the safety lever on the cocking handle is pulled all the way outthen the handle can be pulled to the rear.
4. The weapon is now ready to fire.

I.12   Soviet Dragunov Sniper Rifle (SVD)

TECHNICAL DATA

Caliber: 7.62×54mmR
Operation: Gas, rotating bolt, semiautomatic only
Magnification: 4
Max eff range: 800–1000 m
Max range: 1300 m
Feed: 10 rd box mag
Rate of fire: 30 rds/min
Weight: 9 lbs.
Length: 48 in.
Length, with bayonet: 54 in.
Length, barrel: 24.4 in.
Rifling: 4 right
Rate of twist: 1:10 in.
Land width: 0.072–0.074 in.
Groove width: 0.158–0.160 in.
Extractor: 2 o'clock
Ejector: 9 o'clock
Manufacturer: Izhevsk Mechanical Plant, Soviet Union

Weapons Brief

The Dragunov Sniper Rifle (SVD) has long been the standard-issue sniper rifle utilized by Soviet motorized rifle platoons and former Warsaw Pact militaries. It uses a short-stroke gas piston and the gas chamber has a two-position manual gas regulator, which allows it to use poor-quality ammunition.

The SVD is easily distinguished by its length and by the skeletonized butt stock and telescopic sight (PSO-1). The serial number is located on the bottom of the receiver. It also has a manual safety selector lever, which is located on the right side of the receiver. When the selector lever is manually placed in the upward position, the weapon is on safe.

Operability

1. Visually check that the chamber is clear.
2. Visually check the barrel to ensure that there is no obstruction.
3. Ensure that the weapon is in the safe position by placing the selector on the right side of the receiver in the upper position.
4. Insert the magazine and pull the charging handle to the rear and release it.
5. Place the selector lever in the downward position and begin firing.

I.13 Romanian FPK Sniper Rifle

TECHNICAL DATA

Caliber: 7.62×54mmR, 7.62×51mm
LPS-1 sight power: 4×
Operation: Gas operated, long-stroke semiautomatic
Max eff range: 800 m
Maximum range: 1300 m
Feed: 10 rd box mag
Rate of fire: 30 rds/min
Weight: 9.2 lbs.
Length: 44.5 in.
Length, with bayonet: 50.5 in.
Length, barrel: 25.6 in.
Rifling: 4 right
Rate of twist: 1:12.5 in.
Land width: 0.070–0.074 in.
Groove width: 0.156–0.160 in.
Extractor: 2 o'clock
Ejector: 9 o'clock
Manufacturer: Fabrica de Arme Cugir, SA, Romania

Weapons Brief

The FPK is very similar to the Soviet SVD with its skeletal butt stock and the LPS-1 telescopic sight mounted on the left side of its receiver. The FPK is known as a re-designed RPK light machine gun which chambers the 7.62×54mmR cartridge. The FPK can be easily distinguished from the SVD by its stamped metal receiver, RPK-type selector lever, gas tube, and machined cuts on its muzzle compensator. The serial number is located on the bottom of the receiver. It has a manual safety selector lever located on the right side of the receiver. With the lever placed in the upward position, the weapon is on safe.

There are several variants of the FPK, which includes the popular Velmet M-76 and M-78, which is chambered for four types of ammunition: 7.62×39mm, 5.56×45mm, 7.62×51mm, and 30.06 caliber (7.62×63mm).

Romania is still the primary country that utilizes the FPK sniper rifle, although they have been used throughout the Persian Gulf and Middle East and have been found in large quantities on the battlefield.

Operability

1. Visually check that the chamber is clear.
2. Visually check the barrel to ensure that there is no obstruction.
3. Insert magazine.
4. Move the selector to semiautomatic.
5. Pull the charging handle to the rear and release it.
6. The weapon is now ready to fire.

I.14   Soviet RPD Machine Gun

TECHNICAL DATA

Caliber: 7.62 × 39mm
Operation: Gas operated, open bolt
Max eff range: 800 m
Feed: Belt, from drum
Rate of fire: 700 rds/min (cyclic)
Weight: 15.6 lbs.
Length, overall: 40.08 in.
Length, barrel: 20.05 in.
Rifling: 4 right
Rate of twist: 1:10 in.
Land width: 0.068–0.072 in.*
Groove width: 0.154–0.158 in.*
Extractor: 6 o'clock
Ejector: 12 o'clock
Manufacturer: Izhevsk Mechanical Plant, Soviet Union

* Varies by country

Weapons Brief

The RPD light machine gun is a squad-level weapon used by the Soviet military. The design work was started by Vasily Degtyarev in 1943 in order to make use of the newly developed 7.62 × 39mm cartridge. It was manufactured in large quantities and formed the standard squad automatic for the Russian military. It was also manufactured by Communist China as the Type 56 and Type 56-1 and by North Korea as the Type 62 light machine gun.

The serial number is found in three locations: on top of the top cover, on the bottom of the receiver, which is just forward of the trigger guard, and on the bottom of the slide. The rotary-style safety lever is located on the right side of the receiver near the trigger guard.

Operability

1. Visually check that the chamber is clear.
2. Visually check the barrel to ensure that there is no obstruction.
3. Ensure that the weapon is on safe and cleared.
4. Insert the non-disintegrating, 100-round, segmented belt feed from the left side.
5. Move the selector lever from safe to automatic.
6. Pull the charging handle to the rear (caution: fires from the open bolt position).
7. The weapon is now ready to fire.

I.15   Czechoslovakian VZ52/57 Light Machine Gun

TECHNICAL DATA

Caliber: 7.62 × 39mm
Operation: Gas, selective fire, tilting breechblock
Max eff range: 800 m
Feed: 25 rd box mag or NDLB
Rate of fire:
         Belt 1200 rds/min
         Mag 900 rds/min
Weight: 18 lbs.
Length, overall: 41 in.
Rifling: 4 right
Rate of twist: 1:11 in.
Land width: 0.075–0.079 in.
Groove width: 0.173–0.176 in.
Extractor: 2 o'clock
Ejector: 6 o'clock
Manufacturer: Ceska Zbrojovka, Czechoslovakia

Weapons Brief

The VZ52 originally fired the 7.62mm short cartridge but was later modified to fire the Soviet 7.62×39mm M43 cartridge and is now identified as the VZ52/57.

The weapon is fitted with a quick-change barrel and bipod. The serial number is located on the left side of the receiver and the top cover. The safety is located on the left side of the receiver near the top of the hand grip area.

During the recent conflicts within Afghanistan and Iraq, only small quantities of these weapons were found on the battlefield and those that were seized were in very poor operating condition.

Operability

1. Visually check that the chamber is clear.
2. Visually check the barrel to ensure that there is no obstruction.
3. Insert the magazine and make sure it locks into position.
4. If using belted ammunition, push upward on the top cover while placing the belt of ammunition in the tray between the feed pawls, and close the cover.

I.16   Soviet PK Machine Gun

TECHNICAL DATA

Caliber: 7.62 × 54mmR
Operation: Gas, automatic only (cyclic)
Max eff range: 1000 m
Feed: 100, 200, 250 rd NDLB
Rate of fire: 690–720 rds/min
Weight: 19.8 lbs.
Length, overall: 45.75 in.
Length, barrel: 23.75 in.
Rifling: 4 right
Rate of twist: 1:10.75 in.
Land width: 0.068–0.072 in.
Groove width: 0.154–0.158 in.
Extractor: 3 o'clock
Ejector: 9 o'clock
Manufacturer: Izhevsk Mechanical Plant, Soviet Union

Weapons Brief

The (PK) Pulemyot Kalashnikova) machine gun is a light, gas operated, rotary bolt, belt fed, automatic weapon that fires from the open-bolt position. It has a manual rotary selector safety located on the leftside of the trigger guard. When placed in the rearward position, the weapon is on safe. The original PK machine gun was introduced in 1964 and since then numerous modifications have been made. PK series machine guns are based on the Kalashnikov assault rifle and light machine gun designs.

Operability

1. Visually check that the chamber is clear.
2. Visually check the barrel to ensure that there is no obstruction.
3. Press the cover locking mechanism forward and lift the top cover.
4. Place the ammunition belt on the feed tray ensuring that the first cartridge is in the cartridge holder.
5. Close the feed cover and rotate the safety forward.
6. Pull the charging handle rearward and now move it fully forward.
7. The weapon is now ready to fire.

I.17   Soviet DShK 1938/46 Heavy Machine Gun

TECHNICAL DATA

Caliber: 12.7×108mm
Operation: Gas, automatic only, belt fed
Max range: 2500 m
Max eff range: 2500 m
Feed: 50 rd metallic link belt
Rate of fire: 600 rds/min
Weight, gun: 78.5 lbs.
Weight, barrel: 28 lbs.
Length, gun: 62.5 in.
Length, barrel: 42 in.
Rifling: 8 right
Rate of twist: 1:15.5 in.
Land width: 0.088–0.092 in.
Groove width: 0.111–0.115 in.
Extractor: 2 o'clock
Ejector: 7 o'clock
Manufacturer: Izhevsk Mechanical Factory, Soviet Union

Weapons Brief

The DShK 1938/46 heavy machine gun was adopted by the former Soviet Union in 1938 as a dual antiaircraft and anti-vehicular weapon. It is easily recognized by its large flat and rectangular feed cover. The serial number is located on top of the receiver and on top of the top cover. The rotary-type safety lever is located on the right side of the receiver, near the trigger guard.

The DShK can be found in use by former Warsaw Pact and Asian Communist countries. It's still primarily used as antiaircraft armament for the Soviet tanks, heavy assault guns, and Armored Personnel Carriers (APC), and as electronic armament on the T-10 heavy tank.

During the recent conflicts throughout Afghanistan and Iraq, numerous DShK weapons were seized on the battlefield.

Operability

1. Visually check that the chamber is clear.
2. Visually check the barrel to ensure that there is no obstruction.
3. Lift the top cover.
4. Insert the belt of ammunition into the feed tray area and close the top cover.
5. Pull the charging handle to the rear.
6. Push the trigger when ready to fire.

I.18   Soviet AGS-17 Grenade Launcher

courtesy wiki commons

TECHNICAL DATA

Caliber: 30mm
Operation: Blowback
Max range: 1700 m
Max eff range: Direct 800–1200 m, indirect 1700 m
Feed: 29 rd belt
Rate of fire: 400 rds/min
Weight: 39 lbs.
Weight, with mount and sight: 68 lbs.
Length, overall: 33 in.
Length, barrel: 11 in.
Manufacturer: JSC Konstrukforskoe Buro Priborostroeniya (KBP),
    Russian Defense Industry, based in Tula

Weapons Brief

The AGS-17 was introduced into the Soviet military around 1975 and was found on the battlefields in Afghanistan and throughout the Middle East during recent conflicts.

They are used as the company's base of fire during an advance movement. It is still in use by the Russian Army as a direct fire support weapon for infantry troops; it is also installed in several vehicle mounts and turrets along with machine guns, guided rocket launchers, and sighting equipment.

Operability

1. Visually check that the chamber is clear.
2. Visually check the barrel to ensure that there is no obstruction.
3. Cock the bolt by pulling back on the operating knob.
4. Release the knob, thus allowing the bolt to chamber a cartridge.
5. The AGS-17 is now ready to fire.

I.19   Soviet RPG-7 Antitank Grenade Launcher

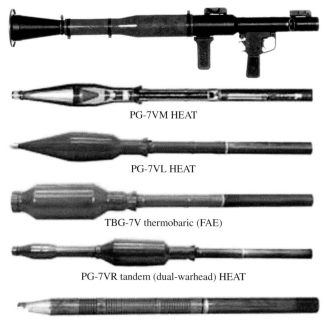

PG-7VM HEAT

PG-7VL HEAT

TBG-7V thermobaric (FAE)

PG-7VR tandem (dual-warhead) HEAT

OG-7V fragmentation antipersonnel

courtesy wiki commons

TECHNICAL DATA

Caliber: 40mm
Operation: Recoilless
Max range: 920 m
Max eff range: Moving 300 m, stationary 500 m
Feed: Single shot, manually loaded
Crew: 2 people
Armor penetration: 500mm
Rate of fire: 4–7 rds/min
Weight: 15 lbs.
Length: 37 in.
Manufacturer: Bazalt, Soviet Union

Weapons Brief

The RPG-7 is a reusable rocket-propelled grenade; a shoulder-fired, muzzle-loaded, smooth-bore, antitank weapon. It fires a variety of fin-stabilized rounds, including a rocket-assisted, high explosive antitank (HEAT) round. The launcher consists of the pistol grip trigger assembly, rear hand grip, sight posts, and optical sight mount used for mounting the PGO-7 optical sight. The serial number is located on the left side of the receiver, just behind the trigger. The safety is a push-button type, which is located directly behind the trigger.

The RPG-7 is still a standard issue weapon in the Soviet military and all former Warsaw Pact countries.

Operability

1. Visually check that the chamber is clear.
2. Visually check the barrel to ensure that there is no obstruction.
3. Ensure that the weapon is on safe; push the safety until it protrudes on the right side of the trigger assembly.
4. Inspect the launcher for any signs of damage.
5. Insert the grenade until the indicator stem of the grenade is fully seated in the notch located at the top of the muzzle end.
6. Adjust sights to the estimated range of the target and make sure the backblast area is clear from 50 to 100 m.
7. Manually cock the hammer while releasing the safety and it's now ready to fire.

I.20   Soviet RPG-18 Antitank Rocket Launcher

TECHNICAL DATA

Caliber: 64mm rocket
Operation: Recoilless
Max eff range: 200 m
Feed: One-shot preloaded, disposable
Armor penetration: 375mm
Rate of fire: 4–7 rds/min
Weight: 6 lbs.
Length, closed: 28 in.
Length, extended: 41 in.
Manufacturer: Bazalt, Soviet

Weapons Brief

The Soviet RPG-18 is a light, inexpensive, short-range, tube-launched, disposable infantry antitank rocket weapon system. The RPG-18 consists of two concentric tubes.

The RPG-18 is a one-shot, unguided, disposable grenade launcher. The RPG-18 is somewhat cumbersome having limited range, accuracy, and effectiveness. Once the tube is extended, the weapon can't be collapsed and therefore *must* be fired.

Operability

1. Visually check that the chamber is clear.
2. Visually check the barrel to ensure that there is no obstruction.
3. Pull the rear pin downward).
4. Extend launcher until it's locked.
5. Place the launcher on your shoulder.
6. Check your back-blast area and ensure you have a safety zone.
7. Select your target and fire.

I.21   Soviet SPG-9 73mm Recoilless Gun

courtesy wiki commons

TECHNICAL DATA

Caliber: 73mm
Operation: Recoilless, manually loaded
Max range: 1200–1500 m
Max eff range: 800 m
Feed: Single shot
Rate of fire: 5–6 rds/min
Weight: 105 lbs.
Weight with tripod: 131 lbs.
Length: 83 in.
Manufacturer: State Factories, Soviet

Weapons Brief

The SPG-9 was first produced in 1968 and is still in use by the former Soviet motorized rifle battalions and airborne troops. The weapon uses ammunition known as the PG-9.

The SPG-9D is the airborne version and has detachable wheels.

Only a few of these weapons were observed on the battlefield during the Afghanistan and Iraq conflicts.

Operability

1. Visually check that the chamber is clear.
2. Visually check the barrel to ensure that there is no obstruction.
3. Unlock the breech.
4. Load the appropriate munition into the chamber.
5. Lock the chamber.
6. Move the trigger mechanism in the rear and engage the target.
7. Pull the trigger when ready to fire.

Foreign and U.S. Weapons Review

This section outlines those weapons commonly encountered. The following information includes detailed images for definitive identification, ammunition types and applications, rate of fire, barrel's rate of twist, effective range, origin of manufacture, and other operational characteristics, as well as physical characteristics such as weight and length of the barrel/overall weapon. Also included is a listing of strengths and weaknesses for each, a brief summary of functionality, and precautionary information.

Contents

Foreign and U.S. Weapons Review Sheet

Symbols

R = rimmed m = meters
rd = round/ctgs
HIP = high-impact plastic
NDLB = non disintegrating link belt

II.1   Foreign Handguns/Sidearms

Weapon	Country	Caliber	Feed	Lock	Operation	Eff Range (m)
P-38	Germany	9×19mm	8 rd box	Hinged lock block	Recoil	50
PA-63	Hungary	9×18mm Makarov	7 rd mag	None	Blowback	50
PM Pistol	Soviet	9×18mm	8 rd box	None	Blowback	50
P7PSP	Germany	9×19mm	8 rd box	None	Blowback	50
TARIQ	Iraq	9×19mm	8 rd box	None	Recoil	50
TT-33	Soviet	7.62×25mm	8 rd box	Locking lugs	Recoil	50
74	Iraq	7.65×17mm	8 rd mag	None	Blowback	50

II.2   Foreign Submachine Guns

Weapon	Country	Caliber	Feed	Lock	Operation	Eff Range (m)
Madsen	Denmark	9×19mm	32 rd box	None	Blowback	200
MAT1	France	9×19mm	34 rd box	None	Blowback	200
MP-5	Germany	9×19mm	15/30 rd box	Roller actuated delay	Delayed blowback	200
MP-40	Germany	9×19mm	32 rd box	None	Blowback	200
PPS-43	Soviet	7.62×25mm Tokarev	35 rd box	None	Blowback	200
PPSh-41	Soviet	7.62×25mm Tokarev	31 rd mag/ 71 drum	None	Blowback	200
STEN	UK	9×19mm	32 rd box	None	Blowback	200
STERLING	UK	9×19mm	34 rd box	None	Blowback	200
UZI	Israel	9×19mm	24/32 rd box	None	Blowback	200

II.3   Foreign Assault Rifles

Weapon	Country	Caliber	Feed	Lock	Operation	Eff Range (m)
AK-47	Soviet	7.62×39mm	30 rd box	Rotary bolt	Gas	300–400 m
AKM	Soviet	7.62×39mm	30 rd box	Rotary bolt	Gas	300–400 m
AK-74	Soviet	5.45×39mm	30 rd HIP	Rotary bolt	Gas	400–500 m
AKS-74	Soviet	5.45×39mm	30 rd HIP	Rotary bolt	Gas	400–500 m
AMD-65	Hungary	7.62×39mm	30 rd box	Rotary bolt	Gas	300–400 m
GALIL	Israel	5.56×45mm	35 rd box	Rotary bolt	Gas	500 m
MPiK	Germany	7.62×39mm	30 rd box	Rotary bolt	Gas	300–400 m
MPiKM	Germany	7.62×39mm	30 rd box	Rotary bolt	Gas	300–400 m
MPiKMS	Germany	7.62×39mm	30 rd box	Rotary bolt	Gas	300–400 m
MPiKS-74	Germany	5.45×39mm	30 rd HIP	Rotary bolt	Gas	400–500 m
M70AB2	Yugoslavia	7.62×39mm	30 rd box	Rotary bolt	Gas	300–400 m
PMK	Poland	7.62×39mm	30 rd box	Rotary bolt	Gas	300–400 m
PMKD-60	Poland	7.62×39mm	30 rd box	Rotary bolt	Gas	300–400 m
TABUK	Iraq	7.62×39mm	30 rd box	Rotary bolt	Gas	300–400 m
Type 56	PRC	7.62×39mm	30 rd box	Rotary bolt	Gas	300–400 m
Type 56-1	PRC	7.62×39mm	30 rd box	Rotary bolt	Gas	300–500 m
Type 68	North Korea	7.62×39mm	30 rd box	Rotary bolt	Gas	300–400 m
VZ-58P	Czech	7.62×39mm	30 rd box	Pivoting block	Gas	300–400 m
VZ-58 V	Czech	7.62×39mm	30 rd box	Pivoting block	Gas	300–400 m

II.4   Foreign Service Rifles

Weapon	Country	Caliber	Feed	Lock	Operation	Eff Range (m)
FN-L1A1	Belgium	7.62×51mm	20 rd box	Tilting block	Gas	600 m
FPK	Romania	7.62×54mmR	10 rd box	Rotary bolt	Gas	800 m
G3	Germany	7.62×51mmR	20 rd box	Roller delayed	Delayed blowback	400 m
M1944	Soviet	7.62×54mmR	5 rd internal	Bolt action	Manual	450 m
SVD	Soviet	7.62×54mmR	10 rd box	Rotary bolt	Gas	1000 m
Type 63	N. Korea	7.62×39mm	10 rd box	Internal tilting block	Gas	450 m
Type 68	PRC	7.62×39mm	15 rd box	Rotary bolt	Gas	400 m

II.5   Foreign Machine Guns

Weapon	Country	Caliber	Feed	Lock	Operation	Eff Range (m)
AL QUDS	Iraq	7.62×39mm	30 rd box	Rotary bolt	Gas	800 m
BREN MK I	UK	.303 cal	30 rd box	Tilting block	Gas	600 m
DShK 38/46	Soviet	12.7×108mm	50 rd NDLB	Locking lugs	Gas	1500 m antiaircraft
FN MAG	Belgium	7.62×51mm	50 rd belt	Locking lever	Gas	1200 m
MG-42	Germany	7.92×57mm	50 rd NDLB	Roller locking	Recoil	1200 m
PK SERIES	Soviet	7.62×54mmR	100/200/250 rd NDLB	Rotary bolt	Gas	1000 m
RPD	Soviet	7.62×39mm	100 rd NDLB	Side lugs	Gas	800 m
RPK	Soviet	7.62×39mm	30/40 rd box/75 rd drum	Rotary bolt	Gas	800 m
RPK-74	Soviet	5.45×39mm	30 rd HIP	Rotary bolt	Gas	800 m

II.6   Foreign Anti-Armor

Weapon	Country	Caliber	Feed	Lock	Operation	Eff Range (m)
AL NASIRAH	Iraq	40×70mm HEAT	Manual	Friction	Manual	300 m moving
		40×85mm HEAT	Manual	Friction	Manual	500 m stationary
		40×40mm AP	Manual	Friction	Manual	
RPG-2	Soviet	40×82mm HEAT	Manual	Friction	Manual	150 m
RPG-7	Soviet	40×70mm HEAT	Manual	Friction	Manual	300 m moving
		40×85mm HEAT	Manual	Friction	Manual	500 m stationary
		40×40mm AP	Manual	Friction	Manual	
SPG-9	Soviet	73mm HEAT	Manual	Breech	Manual	1300 (rocket assist)
Type 69	PRC	40×70mm HEAT	Manual	Friction	Manual	300 m moving
		40×85mm HEAT	Manual	Breech	Manual	
		40×40mm AP	Manual	Breech	Manual	

II.7   Foreign Grenade Launchers

Weapon	Country	Caliber	Feed	Lock	Operation	Eff Range (m)
AGS-17	Soviet	30mm	29 rd NDLB	None	Blowback	700–800 m direct; 1200 m indirect
BG-15	Soviet	40mm	Muzzle fed	Detent	Manual	400 m

II.8   U.S. Service Pistols

Weapon	Country	Caliber	Feed	Lock	Operation	Eff Range (m)
M9/M9A1	U.S.	9mm Luger	15 rd mag	Locking block	Short recoil	50 m
M1911/ M1911A1	U.S.	.45 CP caliber	7 rd mag	Locking lugs	Short recoil	50 m
M11	U.S.	9mm Luger	13 rd mag	Mechanical lock	Short recoil	50 m

II.9   U.S. Grenade Launchers

Weapon	Country	Caliber	Feed	Lock	Operation	Eff Range (m)
Mk 19	U.S.	40mm	Automatic	Belted	Blowback	150 m point target
M79	U.S.	40mm	Single shot	Breech loaded	Pump action	150 m point target
M203	U.S.	40mm	Single shot	Breech loaded	Pump action	1500 m point target

II.10   U.S. Assault Rifles

Weapon	Country	Caliber	Feed	Lock	Operation	Eff Range (m)
M1 Carbine	U.S.	7.62×33mm	15 rd mag	Rotating bolt	Gas	300 m
M4 Carbine	U.S.	5.56×45mm (NATO)	20/30 rd mag	Rotating bolt	Gas	600 m
M16	U.S.	5.56×45mm (NATO)	20/30 rd mag	Rotating bolt	Gas	550 m
M16A1	U.S.	5.56×45mm (NATO)	20/30 rd mag	Rotating bolt	Gas	460 m
M16A2	U.S.	5.56×45mm (NATO)	20/30 rd mag	Rotating bolt	Gas	550 m ind target

II.11   U.S. Service Rifles

Weapon	Country	Caliber	Feed	Lock	Operation	Eff Range (m)
BAR(M1918A2)	U.S.	7.62 × 63mm, .30 cal	20 rd mag	Projecting lug	Air-cooled, gas	550 m
M1 Garand	U.S.	7.62 × 63mm, .30 cal	8 rd en bloc mag	Rotating bolt	Air-cooled, gas	460 m
M14	U.S.	7.62 × 51mm, .30 cal	20 rd mag	Rotating bolt	Air-cooled/ gas	460 m

II.12   U.S. Submachine Guns

Weapon	Country	Caliber	Feed	Lock	Operation	Eff Range (m)
M3	U.S.	.45 ACP cal	30 rd mag	None	Air-cooled, blowback	90 m
M3A1	U.S.	.45 ACP cal	30 rd mag	None	Air-cooled/ blowback	90 m

II.13   U.S. Machine Guns

Weapon	Country	Caliber	Feed	Lock	Operation	Eff Range (m)
M2 HB	U.S.	.50 cal	Belt	Barrel extension	Air-cooled recoil	1830 m
M60	U.S.	7.62×51mm, .30 cal	Belt	Bolt cam surfaces	Gas/open bolt	1100 m
M73	U.S.	7.62×51mm, .30 cal	Belt	Barrel extension	Air-cooled, gas	900 m
M85	U.S.	.50 cal	Belt	Barrel extension	Recoil	1830 m
M240	U.S.	7.62×51mm, .30 cal	Belt	Bolt cam surfaces	Gas/open bolt	900 m
SAW M249	U.S.	5.56×45mm	200 rd box, 20/30 rd mags	Bolt lugs	Gas/recoil	600 m

Notes

Differences M240 Series

M240B—Buttstock, can be ground mounted on M122A1 tripod or bipod

M240G—Buttstock and ground mounted on M122 tripod with flex mount (Marine Corps only)

M240, M240C, M240E1—Spade grip and is pintle mounted, designed as coaxial machine gun for tanks and 7.62mm on light armored vehicles

M240B/M240G—Bipod is integrated into the receiver assembly

M240E5—Aviation variant with spade grip trigger assembly and special mounting for helicopter

Differences M3/M3A1

M3A1—Retracting handle assembly, retracting lever assembly, retracting lever and oiler clip eliminated

Cocking slot spring and oiler cup eliminated

Cocking slot cut top portion of bolt

Ejector groove entire length of the bolt

No more ejection opening and cover assembly

Safety lock further back on cover

Oil reservoir and oiler placed in the pistol grip bracket

Differences M1911/M1911A1

M1911A1 has front sight widened notch and the rear sight was widened to correspond with the front sight

Tang on the grip safety was extended to protect the hand

Mainspring housing was knurled and curved to fit the hand

A clearance cut was made on the receiver for the trigger finger

The face of the trigger was cut back and knurled

Differences M16 Series

M16A1—Forward assist and recesses on bolt carrier

M16A2—Three-round burst, heavier barrel, case deflector for left-handed shooters, round handgrips, two disconnectors for the burst cam, adjustment for windage and elevation on sights and 5/8 in. longer buttstock

M16A3—Essentially the M16A2 with M16A1 fire control group, which was limited to the U.S. Navy

M16A4—Issued to U.S. Marine Corps during Operation Iraqi Freedom, flat-top receiver, handguard with four Picatinny rails for sights and lasers, etc., forward handgrips, removable carrying handle and flashlight capability

Selector Markings and Proofmarks

Selector Settings and Markings

Firearms encountered in the field range from bolt action to fully automatic. Conversely, most of the encountered firearms fire in either the automatic or semiautomatic mode. To facilitate this, the firearms come with some form of selector, usually a lever on the receiver and movable upon the operator's discretion. This lever will usually include a third choice, such as a safety or locked position disabling the operation; the other two positions are semiautomatic, or one shot per trigger pull and automatic, meaning as long as the trigger is held the firearm will continue to fire. The different firearms encountered might well come from different countries of origin with different words for *safe*, *semiautomatic*, and *automatic*, as well as a different "alphabet" to denote those meanings. This section has detailed images and matrices depicting various firearms and selector positions/labeling possibly encountered in theater. Also provided is information describing how these selector markings and proofmarks can be used to identify the country of origin. (Special thanks to Mr. Ryan Coffey, Mr. Aaron Fullerton and Mr. Ed Love).

Proofmarks

Proofmarks are markings stamped into parts of the firearm mechanism, usually the barrel, which are applied after that firearm has various validation tests, such as a proof test. A proof test is the firing of a deliberate overload to test the strength of a firearm barrel and action. Proofmarks may be evaluated to determine the origin of the firearm. The following section includes many proofmark images likely to be encountered in the Persian Gulf, Middle East, and other countries.

Contents

Table III.1   Proofmarks, Selector Markings, and Country of Origin

	Weapon Markings			
		Selector Position		
Factory Proofmark	Safe—Upper	Middle	Lower	Country
⑩		AB	ЕД	Bulgaria
She/✕				Czechoslovakia
①/⑪/⑪/ᶻ₁₄		30	1	Czechoslovakia
☼ 63		D	E	East Germany
◇				East Germany (former)
⋮X⋮ 63		D	E	East Germany (former)
		...	.	Finland
②/⌇₄₅/⌇₄₃		∞	1	Hungary
△ (camel)				Iraq
★ 58ᵓ		러ㅋ ㄴ	다ㅏ ㄴ	North Korea
★/★		러ㅋㄴ	다ㅏㄴ	North Korea
☆ VN				North Vietnam
⑪/(girl) 1962		C	P	Poland
△66 56-1		连	单	PRC
△66		L	D	PRC
m22/m-7		L	D	PRC
△36 ⊏ 六 壽 工		火		PRC
㊷ /△51/☆				PRC
Ⓧ/㊲/△47				PRC
△				PRC
⌒ 216				PRC
△66 出 六 式				PRC

(*Continued*)

Table III.1 (Continued) Proofmarks, Selector Markings, and Country of Origin

	Weapon Markings			
		Selector Position		
Factory Proofmark	Safe—Upper	Middle	Lower	Country
②/△1966/⚠	S	FA	FF	Romania
△ 1954 г.		AB	ОД	USSR
⅃⚡ 1951 г.		AB	ОД	USSR
☆/⊂⊃		ПР	ОГОНЬ	USSR
◇/⊕		ПР	ОГОНЬ	USSR
N R		R	J	Yugoslavia
△ 1954 г.		AB	ОД	USSR

Table III.2 Weapon Selector Markings

Full Auto	Semiauto	Manufacturer	Model
AB	Ед	Bulgaria	AK-47 and AKM
一 天	毕	China	Early production
L	D	China	Type 56 and 56-1
30	1	Czechoslovakia	M58
D	E	East Germany (former)	MPK, MPiKmS MPiKM, MPiKMS
D	E	Egypt	MISR
•••	•	Finland	RYNNAKOKIVAARI (M60 and M62)
∞	1	Hungary	AKM-63 and AMD-65
L	D	Iran	KLS (AKM); KLF (AKMS)
↪	↪	Iraq	Tabuk (AKM)
ㄹㅣㄴ	ㄷㅏㄴ	North Korea	Type 58 (AK-47) and Type 68 (AKM)
C	P	Poland	PMK, PMK-DGN, and KbK AK
FA	FF	Romania	M63; "S" on top for safe position
AB or AV	Од or O	Russia	AK-47, AKS-47, AKM, and AKMS
R	J	Yugoslavia	M64 (AK-47) and M70 (AKM); "U" at top = safe position

Table III.3   Manufacturer Proofmarks

Marking	Producer	Marking	Producer
⑩	Bulgaria	㉑	Bulgaria
㉕	Bulgaria	◡₀	Egypt
◇	East Germany Suhl factory	☼	East Germany
Ⓚ3	East Germany Ernst Thaelman factory	⊗	East Germany
△	Iraq	☆	North Korea
⑪	Poland	△	PRC/China
36 / ⫛	PRC/China	△	PRC/China
㉚₆ Polytec	PRC/China Polytec	66	PRC/China Norinco
㉗ 47	PRC/China	57 57 ☆2	PRC/China
△	Romania	⊕	Russia
▯	Russia Izhevsk factory	△ △	Russia Izhevsk factory
☆ ☆	Russia Tula arsenal	△	Russia Tula arsenal
☆	Russia Polyarny arsenal		

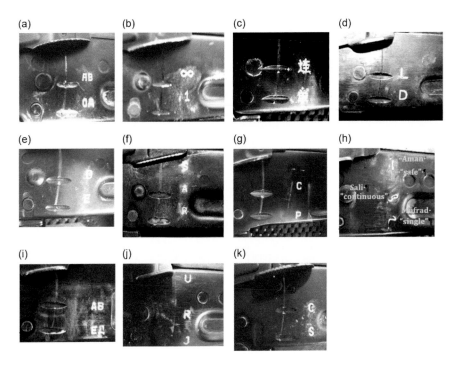

Figure III.1 Assault rifle selector markings – AK (a) Russian AKM, (b) Hungarian AMD-63, (c) Chinese Type 56, domestic, (d) Chinese Type 56, export, (e) East German MPi-KM, (f) Romanian AIM, (g) Polish Kbk AKM, (h) Iraq Tabuk, (i) Bulgarian AKKM, (j) Yugoslavian M70, and (k) East German MPi-69 (.22 caliber AK training). (Note: Special thanks to Aaron Fullerton.)

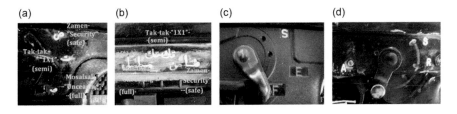

Figure III.2 Other assault rifle selector markings: (a) Iranian G3, (b) Iranian UZI, (c) H&K MP5, and (d) FN FAL.

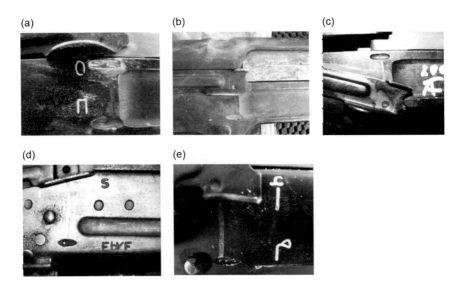

Figure III.3 Semiauto selector markings: (a) Russian SVD (Dragunov), (b) Russian SVD (Dragunov)—alternate, (c) Chinese (Norinco) Type 79, (d) Romanian FPK, and (e) Iraqi Al-Kadissiya.

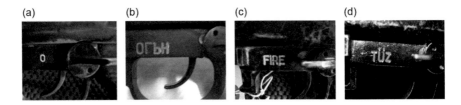

Figure III.4 Machine gun selector markings: (a) Russian PKM, (b) Bulgarian PKM, (c) PKM, and (d) Hungarian PKM.

Armor-Piercing Penetrators Found in the Persian Gulf/ Middle East

Penetrators

A penetrator is a hardened metal core, usually some steel alloy component, within a jacketed bullet designed to improve penetration in "hard" targets. The hardened core of the penetrator is designed to penetrate and disable machinery, vehicles, boats, and aircraft. Although a projectile has a steel core, that steel core would not be classified as a penetrator if the core was mild steel.

Many fabrications of projectiles use a mild steel core because the raw material is more available or economic. Any steel core bullets are still copper or bimetal jacketed so that the barrel of the firearm is not directly exposed to the core.

This section has various images of actual steel cores and penetrators that are likely to be encountered, listing various quantifiable characteristics such as weight, length, shape, and possible associated parts and functionalities, such as base cones and incendiary types, respectively. Also listed are the likely national origins of the penetrators, although in some cases exact duplication makes that determination difficult or impossible.

The main distinction between mild steel cores and penetrators is the hardness of the material. Since many calibers are used in the battlefield, many of these calibers have ammunition loaded with steel cores. Since steel is harder than the traditional lead core material and is far more likely to survive impact relatively intact, it is possible to encounter many different sizes and designs of metal cores. (Special thanks to Mr. Aaron Fullerton.)

Contents

IV.1   7.62×39mm Penetrators

The cores in 7.62×39mm ball ammunition cannot properly be called pen-etrators, but are still sometimes found at incidents with their copper jackets missing. The ball ammunition cores are much milder steel than armor-piercing (AP) penetrators and are often found damaged or deformed after striking material where actual penetrators would still be intact. Most of the cores observed had a flat point nose, making them easy to differentiate from AP penetrators.

As can be seen by the data, there is no significant difference between AP incendiary penetrators in 7.62 × 39mm Russian caliber. It would be virtually impossible to determine the country of origin of an unknown penetrator based on measurements alone.

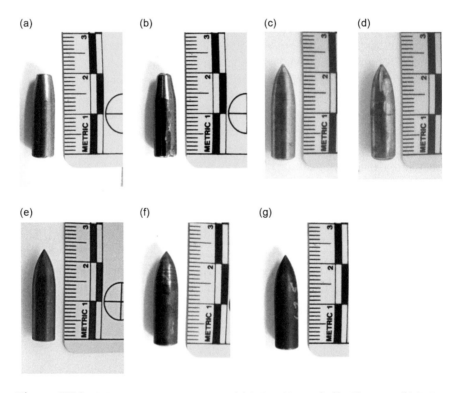

Figure IV.1 7.62×39mm penetrators: (a) 7.62×39mm ball, Chinese, (b) 7.62×39mm ball, Polish, (c) 7.62×39mm API, Hungarian, domestic, (d) 7.62×39mm API, Egyptian, (e) 7.62×39mm API, Chinese, (f) 7.62×39mm API, Iraqi, and (g) 7.62×39mm API, Russian.

Table IV.1   7.62×39mm Specifications and Comparison

	Length (mm)	Shaft Diameter (mm)	Base Diameter (mm)	Boattail Length (mm)	Weight (grains)	Additional Information
Chinese ball	19.7	5.7	5.7	0	55.4	Flat nose
Polish ball	19.9	5.8	5.8	0	54.9	Flat nose
Chinese API	22.4	6.0	6.0	0	61.6	
Egyptian API	22.3	6.1	6.1	0	62.0	
Hungarian API	22.4	6.1	6.1	0	62.2	
Iraqi API	22.3	6.1	6.1	0	62.9	
Russian API	22.5	6.1	6.1	0	62.5	

IV.2   7.62×51mm (.308 Winchester) Penetrators

The cup in the .308 Winchester Swedish AP ammunition is very distinctive, and no other cup of this type was observed in any AP ammunition of any caliber. In the unfired AP bullet, the base of the penetrator sits inside of the cup (Figure 4.2).

IV.3   7.62×54mmR Penetrators

The cores in 7.62 × 54mmR ball ammunition cannot be properly called penetrators, but are still sometimes found at incidents with their copper jackets missing. The ball ammunition cores are much milder steel than AP penetrators and are often found damaged or deformed after hitting material where actual penetrators would still be intact. Most of the cores had a flat point nose, making them easy to differentiate from AP penetrators.

Figure IV.2   308 Winchester Swedish AP.

Table IV.2 308 Winchester Swedish AP Specifications

	Length (mm)	Shaft Diameter (mm)	Base Diameter (mm)	Boattail Length (mm)	Weight (grains)	Remarks
Swedish AP	20.2	5.6	4.8	2.1	91.2	Base of penetrator in cup
Cup	10.1	6.8	6.8	0	6.4	Groove in middle

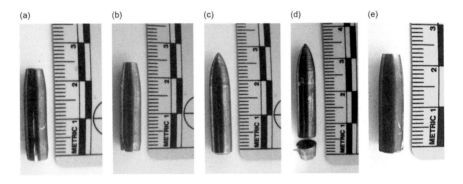

Figure IV.3 7.62×54R penetrators: (a) 7.62×54R Russian LB, (b) 7.62×54R Russian ball, (c) 7.62×54R Chinese API, (d) 7.62×54R Russian API, and (e) 7.62×54R Chinese ball.

Table IV.3 7.62×54R Specifications and Comparison

	Length (mm)	Shaft Diameter (mm)	Bass Diameter (mm)	Boattail Length (mm)	Weight (grains)	Additional Information
Russian light ball	23.2	6.0	5.8	2.8	73.3	Flat point nose
Russian ball	23.4	6.0	5.7	2.8	72.3	Flat point nose
Chinese API	28.3	6.1	6.1	0	82.8	
Russian API	28.4	6.1	6.1	0	83.1	Cup in base
Cup	4.4	6.6	5.5	4.4	*	
Chinese ball	22.5	6.1	6.1	0	71.7	Flat point nose

* Accurate weight could not be determined due to the cup still being filled with chemical.

The cores from the Chinese ball ammunition could be differentiated from the Russian ball ammunition by the lack of a boattail.

The country of origin of penetrators from 7.62×54mmR caliber AP ammunition cannot be determined from measurements of the penetrator alone. The only indicator of the country of manufacture is if the copper cup is found at the scene. The presence of a copper cup indicates that the 7.62×54mmR caliber Armor Piercing Incendiary (API) bullet was manufactured in Russia. The copper cup in unfired ammunition is filled with incendiary compound with the base of the penetrator sitting on top of the cup.

IV.4   .50 Caliber Penetrators

The three types of AP ammunition in the 50 Browning Machine Gun (50 BMG) typically observed are fairly easily differentiated from one another. The core in the 50 BMG caliber ball ammunition has a distinctive "wasp waist" or narrower area on the shaft that corresponds with the cannelure in the copper jacket and a solid base. The 50 BMG caliber API penetrator has a straight shaft and a solid base. The 50 BMG caliber Armor Piercing Incendiary Tracer penetrator externally appears similar to the Armor Piercing Incendiary penetrator but differs in that the base is drilled out and hollow in order to contain the tracer compound. Also, the unfired Armor Piercing Incendiary Tracer bullet has a copper cup in the base in which the penetrator sits. If found to be the proper dimensions, this cup at a scene would indicate the presence of a 50 BMG Armor Piercing Incendiary Tracer bullet, in the absence of a jacket or penetrator.

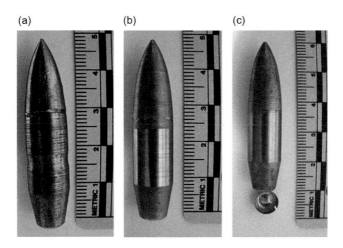

Figure IV.4   .50 caliber penetrators: (a) U.S. .50 BMG ball, (b) U.S. .50 BMG API, and (c) U.S. .50 BMG API-T.

Table IV.4   7.62×54R Specifications and Comparison

	Length (mm)	Shaft Diameter (mm)	Base Diameter (mm)	Boattail Length (mm)	Weight (grains)	Remarks
U.S. ball	47.2	10.8	7.9	10.2	390.1	"Wasp waist"
U.S. API	46.8	10.8	7.8	10.6	393.1	
U.S. API-T	46.7	10.8	7.8	9.8	365.6	Hole in base; base of penetrator in cup
Cup	2.8	8.4	6.4	2.8	1.6	

IV.5   12.7×108mm Penetrators

As can be seen by the data, there is no significant difference between AP incendiary penetrators in 12.7 × 108mm caliber. It would be virtually impossible to determine the country of manufacture of an unknown penetrator off of measurements alone. However, differentiating between API and API-T in 12.7 × 108mm caliber can be easily accomplished. Penetrators in API-T ammunition are considerably shorter to accommodate the tracer material.

(a)   (b)   (c)

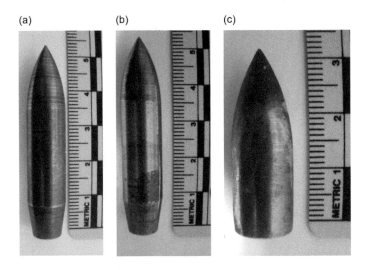

Figure IV.5 12.7×108mm penetrators: (a) 12.7×108mm Chinese API, (b) 12.7×108mm 310 API, and (c) 12.7×108mm API-T.

Table IV.5   12.7 × 108mm Penetrator Specifications and Comparison

	Length (mm)	Shaft Diameter (mm)	Base Diameter (mm)	Boattail Length (mm)	Weight (grains)	Remarks
Chinese API	51.8	10.8	8.3	9.9	459.9	None
310 API	51.6	10.8	8.3	9.9	456.2	None
API-T	31.3	10.8	10.8	0	253.6	None

IV.6   14.4 × 114mm Penetrators

There is no reasonable way to differentiate between the Russian and Romanian API penetrators in 14 × 114mm. This is expected due to most of the Eastern Bloc countries closely following Soviet specifications in ammunition manufacture. The Chinese API penetrator, on the other hand, can be distinguished from the other two based on the difference in the size of the boattail.

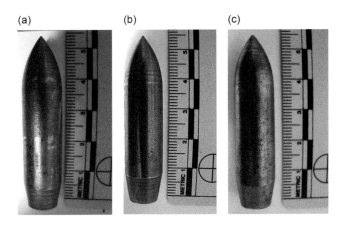

Figure IV.6   12.7 × 108mm penetrators: (a) 14.4 × 114mm Russian API, (b) 14.4 × 114mm Chinese API, and (c) 14.4 × 114mm Romanian API.

Table IV.6   14.4 × 114mm Penetrator Specifications and Comparison

Length (mm)	Length (mm)	Shaft Diameter (mm)	Base Diameter (mm)	Boattail Length (mm)	Weight (grains)	Remarks
Russian API	52.7	12.4	9.9	8.0	628.3	None
Chinese API	52.7	12.4	9.7	9.6	625.6	None
Romanian API	52.5	12.4	10.0	8.0	627.9	None

Headstamps

V

A headstamp is composed of numerals, letters, and symbols (or a combination thereof) stamped into the head of the cartridge case or shotshell to identify the manufacturer, caliber, or gauge or to give additional information. It usually tells who marketed the case. If it is military, it may have the year of manufacture added.

Nations with a defense industry in the interest of national security or economic benefit, or both, have their own munitions plants that produce ammunition for calibers based on the choice of caliber and type of military small arms for defense. Different militaries usually have a core selection of weapons that they produce and/or use; this provides a commonality of ammunition to manufacture, a commonality of firearms to train with, and other associated logistics. That ammunition, since it is likely used by the nationality producing it, has the information written in the language of origin.

The use of headstamps for identification is an important factor in attempting to determine who or what forces might have been involved in a firefight or responsible for a shooting and what caliber and type of firearms were used. Many of the more commonly encountered calibers and headstamps within the Persian Gulf, Middle East, and some other countries are listed on the following pages.

(Special thanks to Mr. Ryan Coffey, Mr. George Kass, and Mr. Ed Love.)

Contents

V.1   7.62×39mm—Egyptian

V.2 7.62×39mm—Iranian

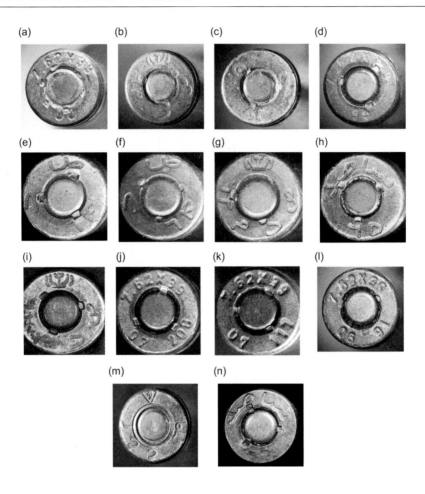

(a)   (b)   (c)   (d)

(e)   (f)   (g)   (h)

(i)   (j)   (k)   (l)

(m)   (n)

V.3 7.62×39mm—Iraqi

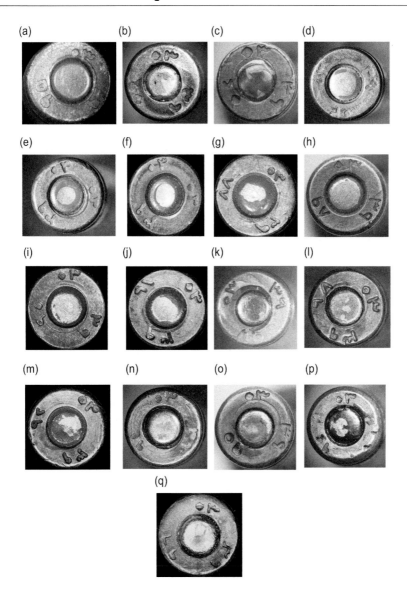

(a) (b) (c) (d)

(e) (f) (g) (h)

(i) (j) (k) (l)

(m) (n) (o) (p)

(q)

V.4   7.62×39mm—Syrian

(a)   (b)

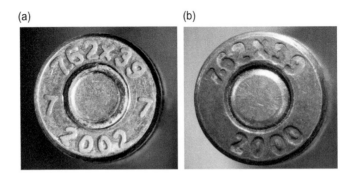

V.5   7.62×39mm—Polish

V.6   7.62×39mm—Bosnian

V.7   7.62×54mmR—Albanian

(a)   (b)   (c)

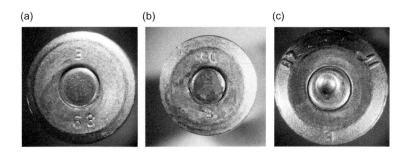

V.8   7.62×54mmR—Bulgarian

V.9   7.62×54mmR—Chinese

V.10   7.62×54mmR—Greek

V.11   7.62×54mmR—Iranian

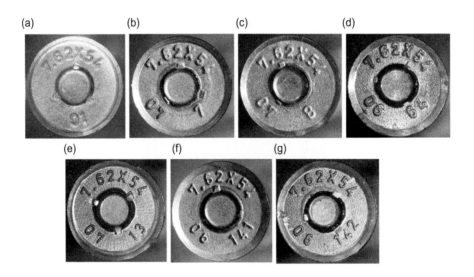

V.12   7.62×54mmR—Romanian

V.13   7.62×54mmR—Iraqi

V.14   7.62×54mmR—Yugoslavian

V.15   7.62×54mmR—Russian

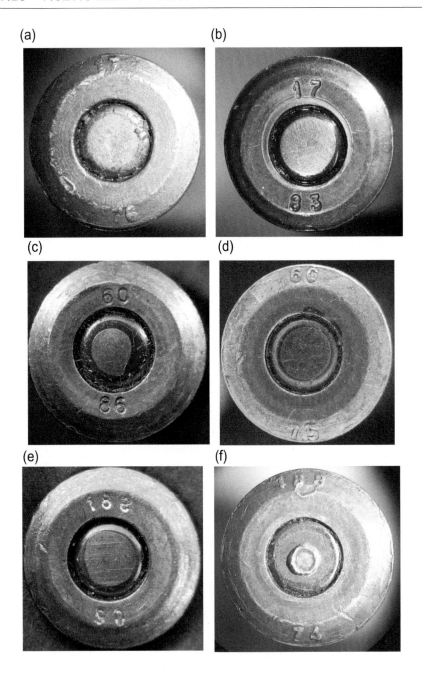

V.16   9×19mm—Iranian

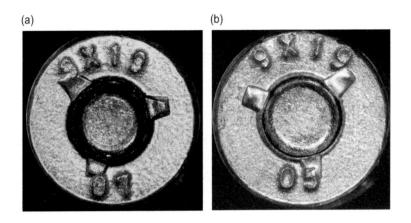

V.17   9×19mm—Iraqi

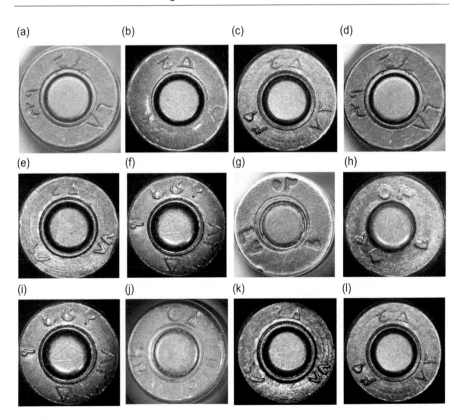

V.18   9×19mm—Romanian

V.19   9×19mm—Czech

V.20   7.62×51mm—U.S.

V.21   7.62×51mm—Swedish

V.22   7.62×51mm—Brazilian

V.23   7.62×51mm—Iranian

V.24   7.62×51mm—Czech

V.25   7.62×51mm—Belgian

V.26   7.62×51mm—Pakistani

V.27   8mm Mauser—Iranian, early

V.28 5.56×45mm—Iranian

V.29 .303 Caliber—Iraq Republic

V.30   Kirkee Indian Arsenal

V.31   12 Gauge—Syrian

V.32   12.7mm—Unknown

V.33   .338—Swiss

Forensic Firearms Training Seminars, Inc. (FFTS), is a provider of a variety of forensic services for attorneys, law enforcement agencies, insurance companies, private investigators, college students, military organizations, and civilian personnel.

Information and inquiries related to this book can be sent to:
Don Mikko, Editor
Forensic Firearms Training Seminars, Inc.
2892 Rex Road
Ellenwood, GA 30294
Fax: (678) 884-5597
Mobile: (678) 371-8478
E-mail: mikko@bellsouth.net
Website: mikkoforensics.com

Bailey Firearm Forensics, Inc. (BFF) is a provider of a variety of forensic services including, but not limited to, shell casing/projectile matching, distance determination, serial number recovery, firearms-related crime scene reconstruction and determination of possible originating firearm, as well as consultant regarding expert witness testimony.

William Bailey
Bailey Forensic Firearms, Inc.
8902 Charlotte Mountain Road
Rougemont, NC 27572
Mobile: (919) 257-0376
E-mail: wbwbailey155@gmail.com

Index

BLUEPRINTS
History
Key Stage 1
Copymasters

Wendy Clemson

Stanley Thornes (Publishers) Ltd

BLUEPRINTS – HOW TO GET MORE INFORMATION

Blueprints is an expanding series of practical teacher's ideas books and photocopiable resources for use in primary schools. Books are available for every Key Stage of every core and foundation subject, as well as for an ever widening range of other primary needs. **Blueprints** are carefully structured around the demands of National Curriculum but may be used successfully by schools and teachers not following the National Curriculum in England and Wales.

Blueprints provide:

- Total National Curriculum coverage
- Hundreds of practical ideas
- Books specifically for the Key Stage you teach
- Flexible resources for the whole school or for individual teachers
- Excellent photocopiable sheets – ideal for assessment, SATs and children's work profiles
- Supreme value.

Books may be bought by credit card over the telephone and information obtained on (0242) 228888. Alternatively, photocopy and return this FREEPOST form to join our mailing list. We will mail you regularly with information on new and existing titles.

Please add my name to the BLUEPRINTS mailing list. *Photocopiable*

Name _____

Address_____

_____

_____

Postcode_____

To: Marketing Services Dept., Stanley Thornes Publishers, FREEPOST (GR 782), Cheltenham, Glos. GL53 1BR

First published in 1992 by:
Stanley Thornes (Publishers) Ltd
Ellenborough House
Wellington Street
CHELTENHAM GL50 1YD

Reprinted 1992
Reprinted 1993

Typeset by Tech-Set, Gateshead, Tyne and Wear.
Printed in Great Britain at The Bath Press, Avon.

British Library Cataloguing in Publication Data

Clemson, Wendy
  Blueprints: history – key stage 1: Copymasters
  – (Blueprints)
  I. Title   II. Series
  372.89

ISBN 07487–1347–6

CONTENTS

INTRODUCTION

In this book there are 96 photocopiable copymasters linked to many of the activities in the Teacher's Resource Book. Where the copymasters are referred to in the text of the Teacher's Resource Book there are instructions on how to use them. They are referred to by number in the Teacher's Resource Book by this symbol The copymasters reinforce and extend activities in the Teacher's Resource Book and provide opportunities to record activities and results in an organised way. When the children have completed these copymasters they can be added to work files or used as exemplar material in pupil profiles. You may also wish to use completed copymasters as a resource for your assessments. There are two record sheets at the back of this book, one for the children's own self-appraisal and one for you to make your comments on their National Curriculum experience.

The copymasters are organised into fifteen key infants topics which are covered in depth in the Teacher's Resource Book. Sheets 79–96 develop general historical skills for Key Stage 1. You will find explanation of how to use them at the front of the Teacher's Resource Book.

Family tree

◯ :draw face 　　▭ :write name

My grandparents

My parents

Me

Happy families

Happy families

Do you think families often had/have these things **in Victorian times** or **now**?

The first one is done for you.

~~Victorian~~
~~Now~~
washing machine

Victorian
Now
quill pens

Victorian
Now
oil lamps

Victorian
Now
car

Victorian
Now
trainers

Victorian
Now
radio

Victorian
Now
dinners

Victorian
Now
candles

Victorian
Now
family Bible

Victorian
Now
TV

Victorian
Now
top hats

Victorian
Now
holidays abroad

Victorian
Now
open fire

Victorian
Now
electric iron

Victorian
Now
copper kettle

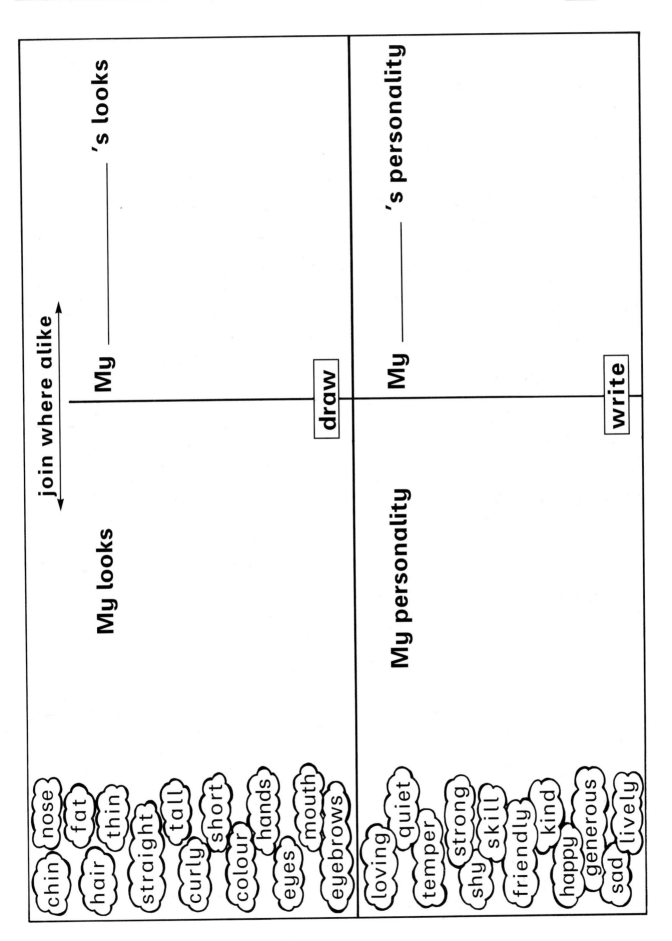

join where alike

My _____ 's looks

My looks

draw

My _____ 's personality

My personality

write

chin · nose · fat · hair · thin · straight · tall · curly · short · colour · hands · eyes · mouth · eyebrows

loving · quiet · temper · strong · shy · skill · friendly · kind · happy · generous · sad · lively

Homes timeline

Colour. Cut out. Put in order.

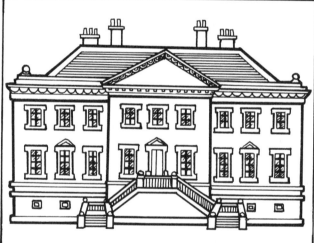

Doors

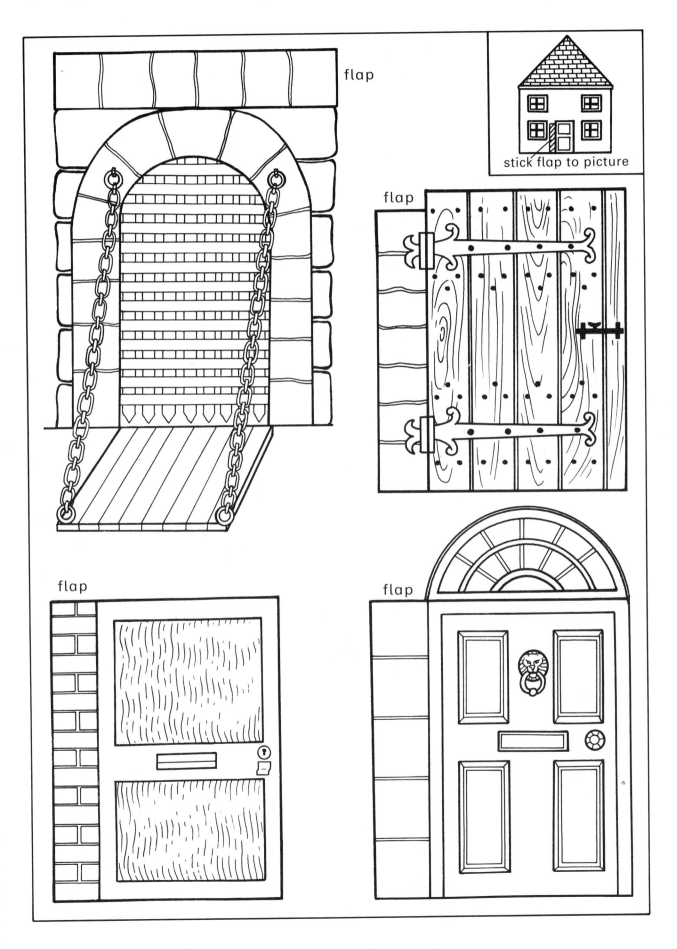

flap

stick flap to picture

flap

flap

flap

My living room
(draw or write)

Doing the washing

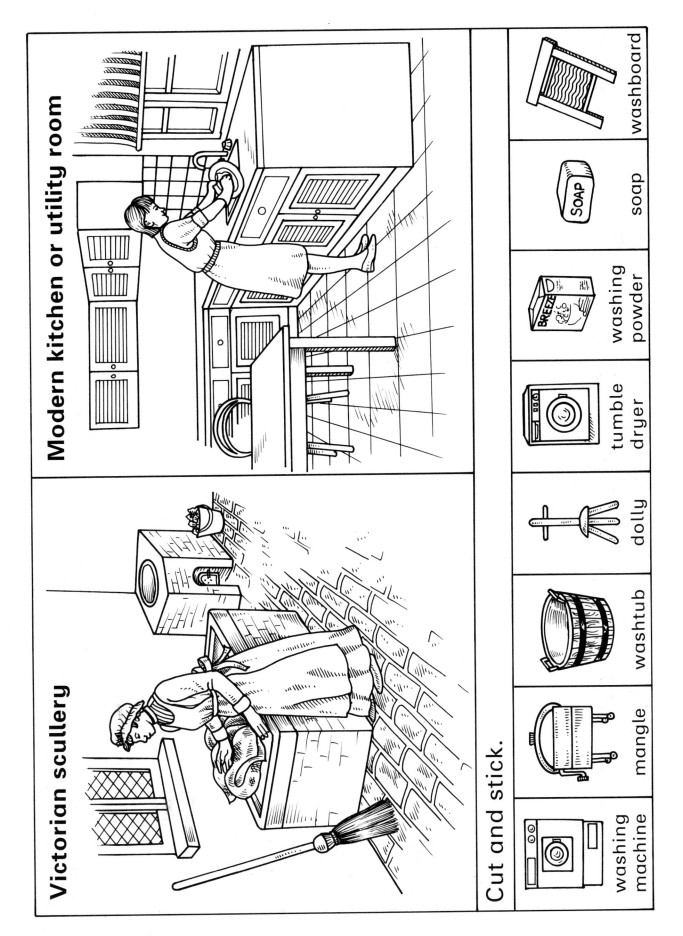

Modern kitchen or utility room

Victorian scullery

Cut and stick.

washboard	soap	washing powder	tumble dryer	dolly	washtub	mangle	washing machine

Museum story

Write the story.

My garden

How my garden looks.

Things that grow in my garden.

What my garden is for.

Picture writing

My code

Picture	(eye)				
Meaning	I/me				

Picture					
Meaning					

My code message

My seal (draw)

How I made my seal.

What a seal is for.

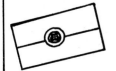 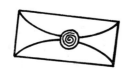 

Cut out. Put in order.

Telling.

Send it by letter post.

Put it on the mail coach.

Send it with a messenger on horseback.

Fax it.

Keeping a diary

My diary for _____ **in** _____

Mondays, Tuesdays, Wednesdays,   (month)
Thursdays, Fridays

Date _____

Date _____

Date _____

Date _____

Date _____

Illuminated initials

Add your initial to each pattern.
Finish and colour the patterns.

Make your own initial patterns here.

Wearing clothes

Why I wear clothes
(draw)

my warm clothes	**clothes that keep me cool**
clothes that keep me safe	**clothes that show I belong to a group**

Fabric sort

Sort the fabrics into sets.

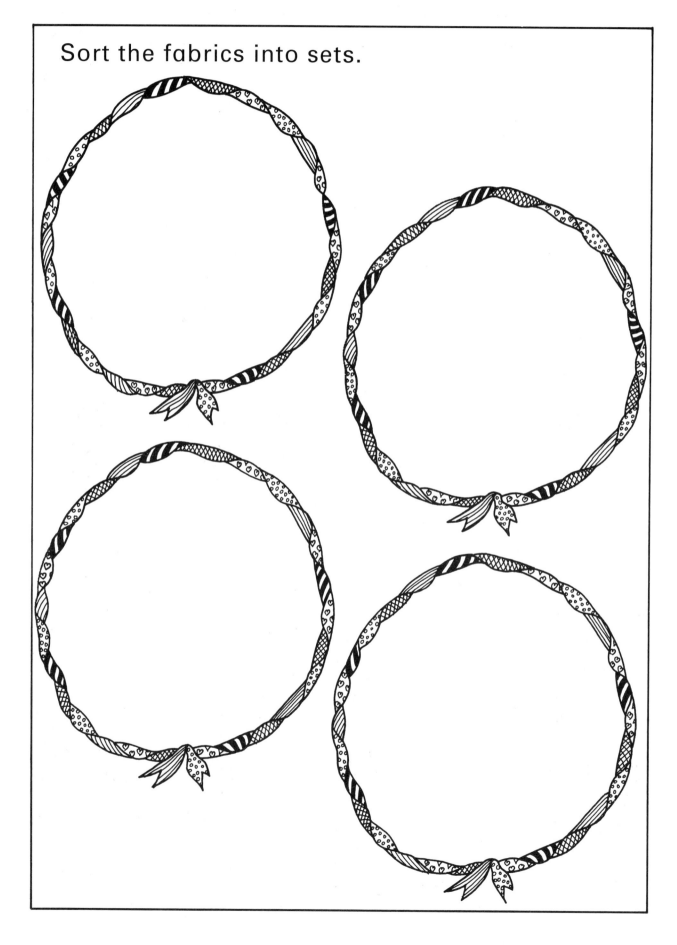

Work clothes

Which toys belong here?

Ring the toys that do not belong
because they had not been invented.

A shoebox theatre

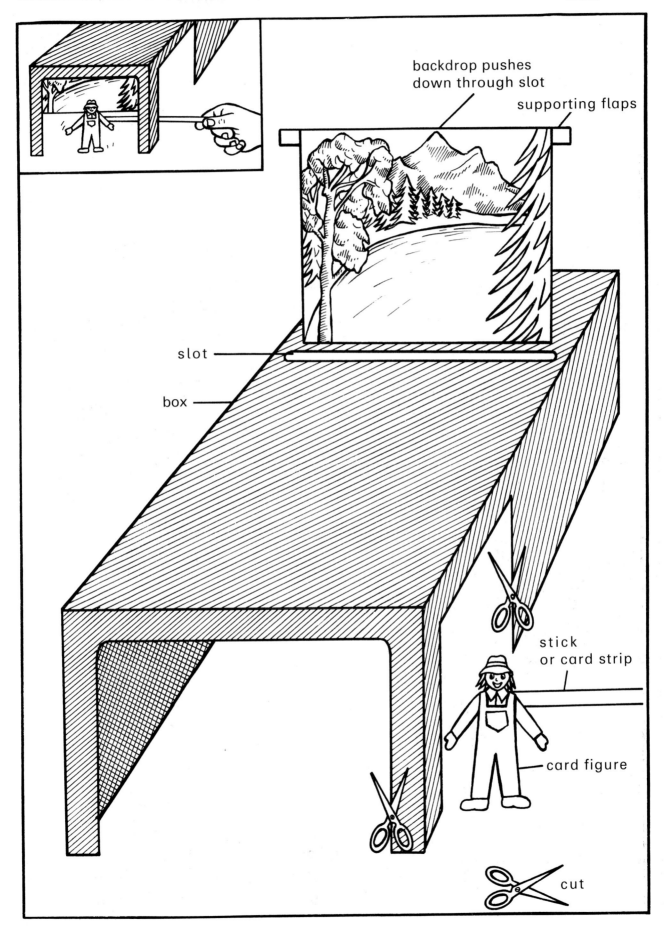

backdrop pushes
down through slot

supporting flaps

slot

box

stick
or card strip

card figure

cut

A paper puppet

Stick on to card. Cut out.
Fix together with 4 paper-fasteners.

a paper-fastener

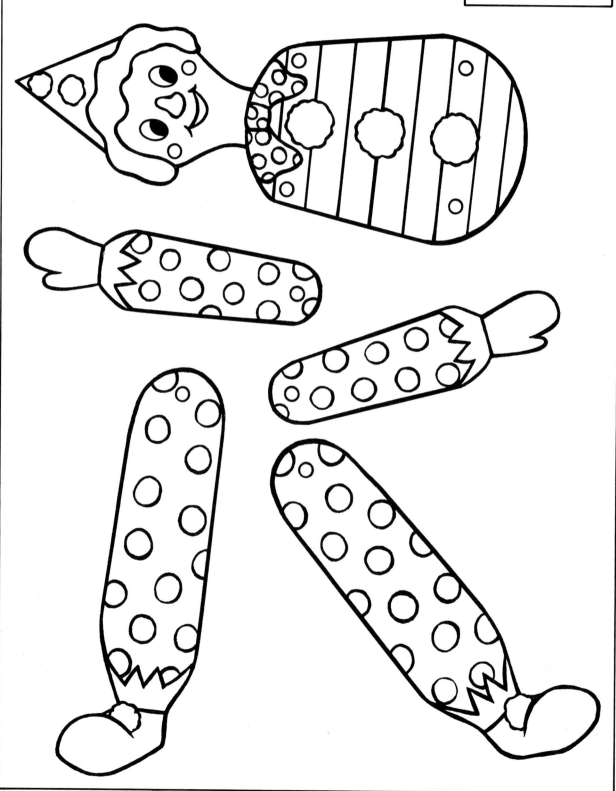

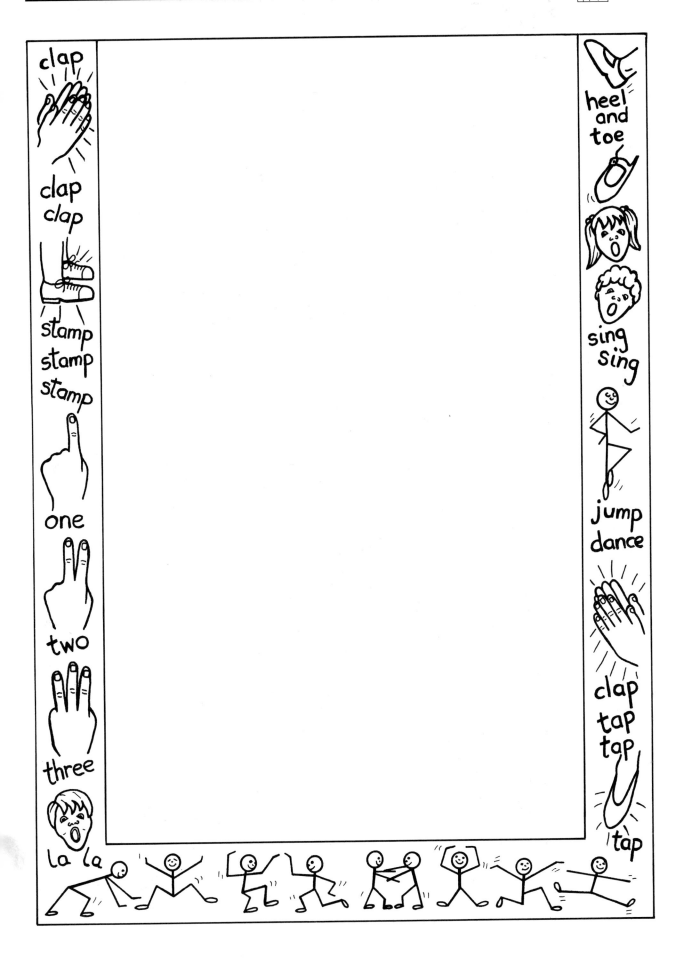

Games long ago

My seaside souvenir

Draw.

What is it? _____

How old is it? _____

Where does it come from? _____

Tell its story.

Draw your own postcard

Come to sunny

Postcard designed by

Postcard timeline

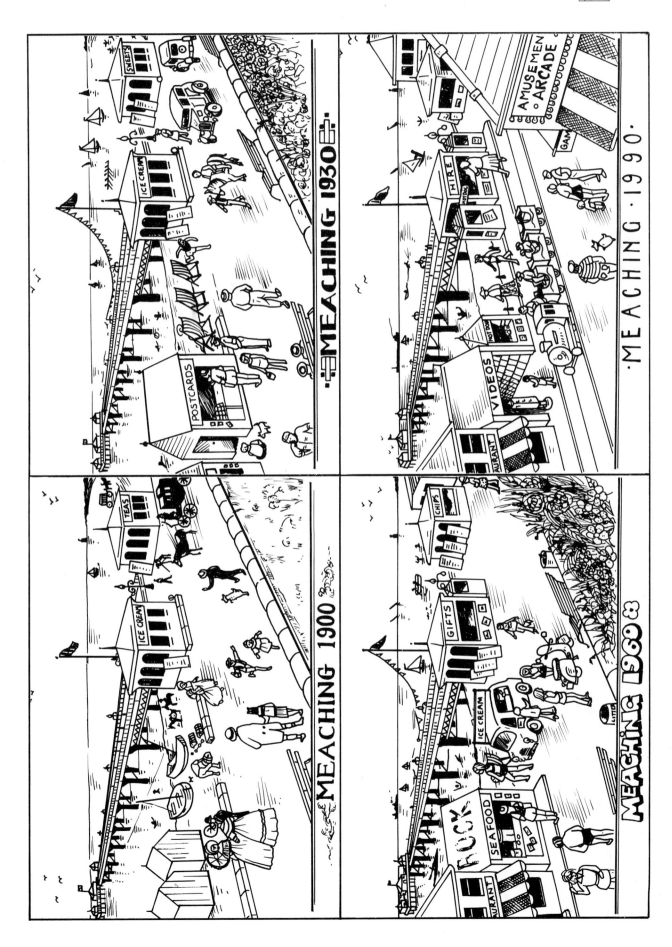

Swimwear

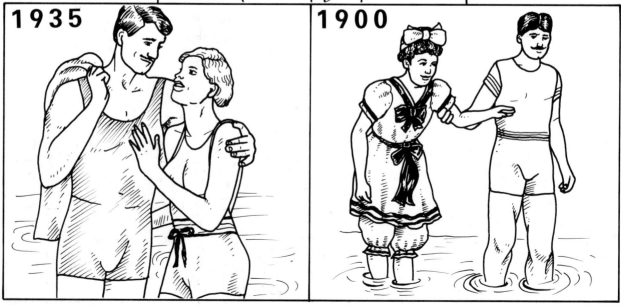

A seaside resort

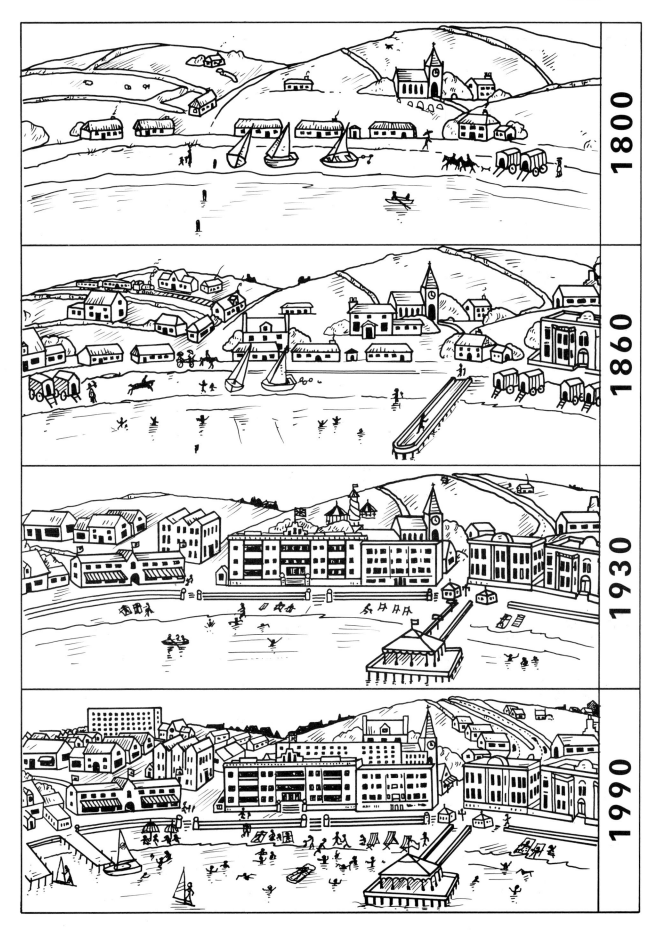

1800

1860

1930

1990

My teacher

What my teacher says.

Things my teacher does in class.

My teacher looks like this.

All about my school

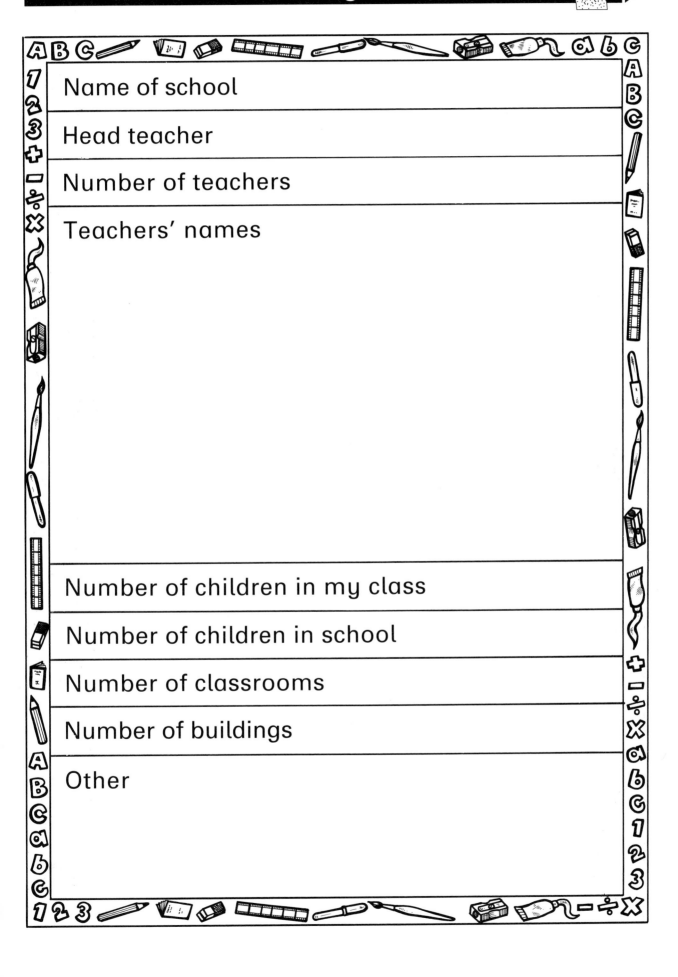

Name of school

Head teacher

Number of teachers

Teachers' names

Number of children in my class

Number of children in school

Number of classrooms

Number of buildings

Other

Grown-ups' schooldays

Did boys and girls do the same work?

What good things happened in school?

What were school dinners like?

What happened when you were naughty?

What did your school look like?

What did you do at school?

What happened when you did good work?

When did school start?

What was school like?

Was there a playground?

A story-book school

My school

The school in the story

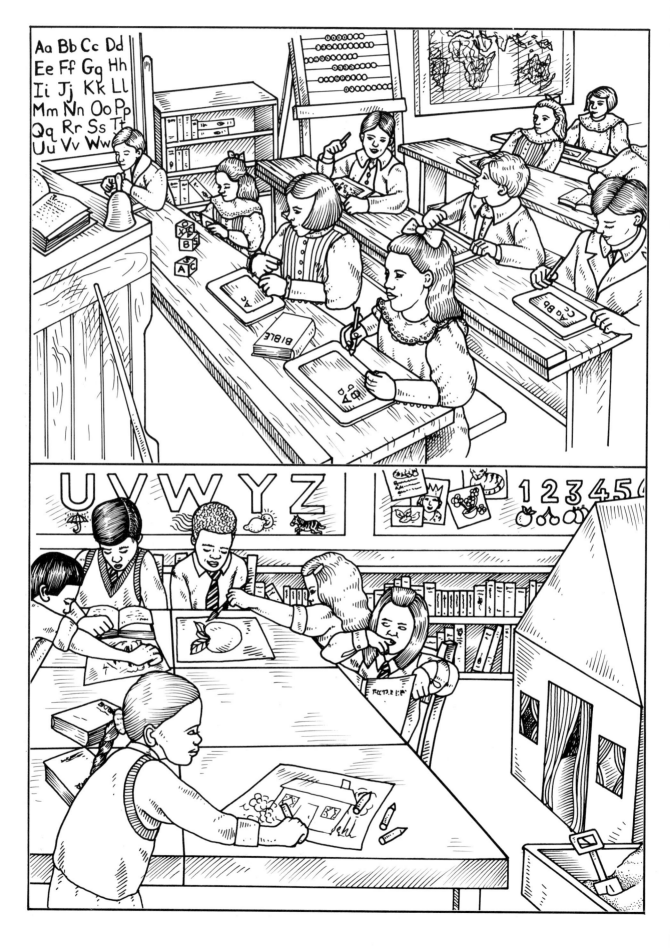

Clothes for school

Edwardian schoolchildren

Schoolchildren today

Mum's or Dad's childhood

I talked to my ——————————.

I asked about these things.

Favourite clothes

Friends

Games

Evenings

Pocket-money

TV

Other things

My Mum/Dad and me

My Mum's or Dad's childhood	Me as a child now

Looking back

In our great-grandparents' time

In our grandparents' time

In our parents' time

My picture story

Draw pictures to tell a story.

What is the warning in this story?

All about food

My favourite foods

Food-tasting record

Name of food	Looks	Taste	Texture	Star rating

Cooking in the past

Cut out the pictures. Put them in order.

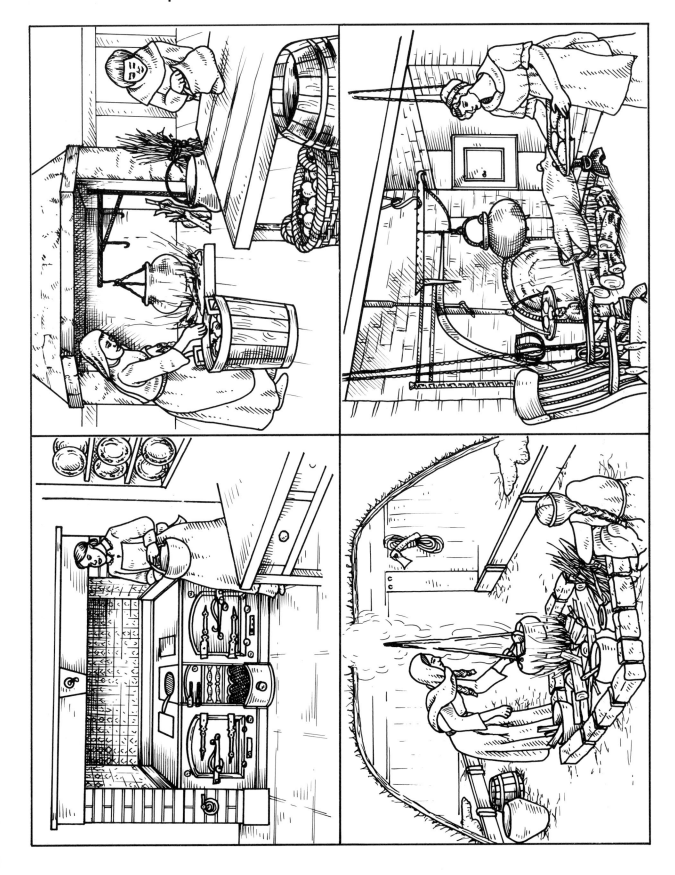

Cutlery

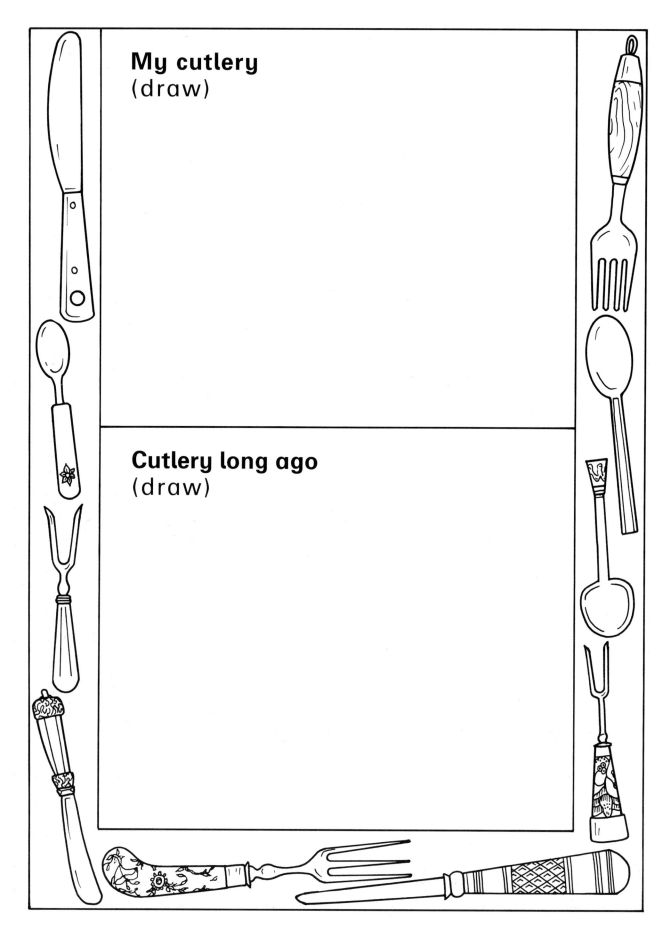

My cutlery
(draw)

Cutlery long ago
(draw)

Meals through the ages

C45

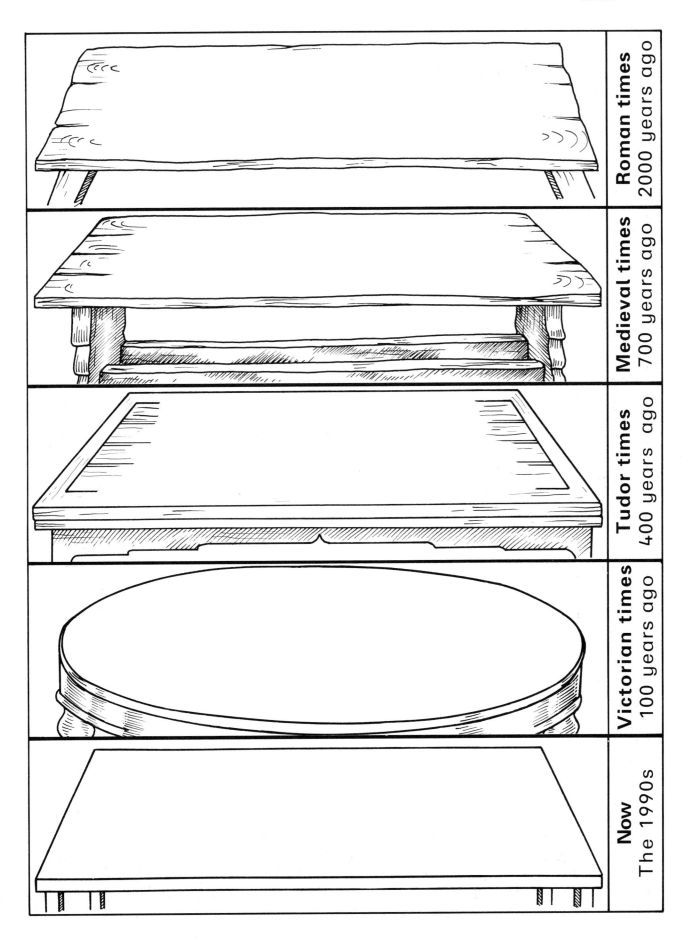

Roman times
2000 years ago

Medieval times
700 years ago

Tudor times
400 years ago

Victorian times
100 years ago

Now
The 1990s

The family car

Our car
(draw)

Make and model _____

More about our car

Family memories of transport

People I talked to	Their memories

What to ask

Have you ever been on a steam train? A trolleybus? A tram?
Tell me about the traffic when you were young.
How did you travel long ago? Where did you go?

Road transport in history

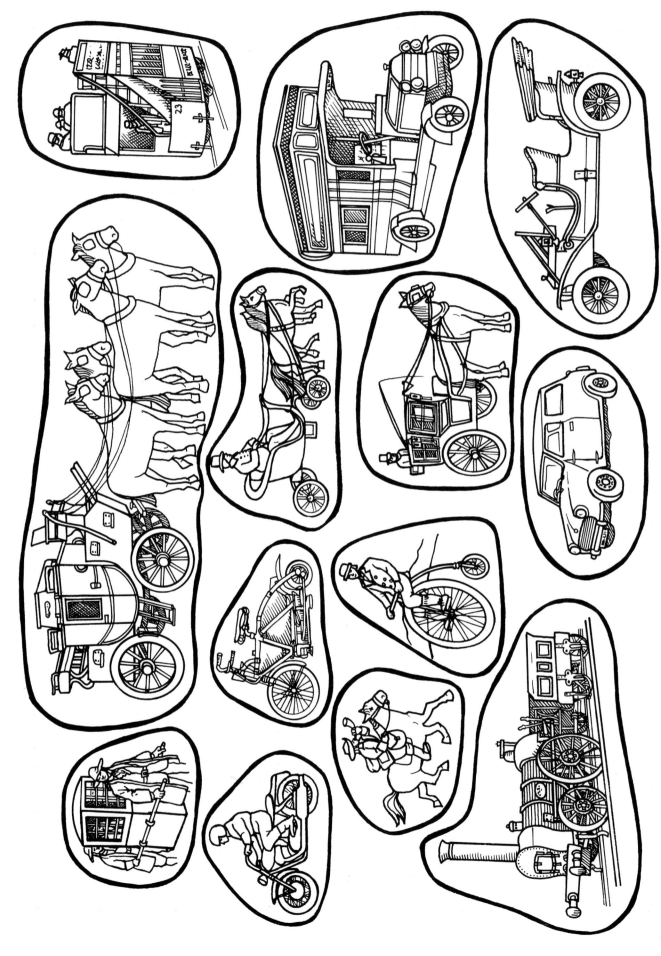

Signposts

A signpost (draw and write)

On my signpost, the nearest place is _____

It is _____ miles away.

How long would it take to get there

on foot? _____

on horseback? _____

by bike? _____

by car? _____

Boats and ships

Draw some boats and ships on the water.

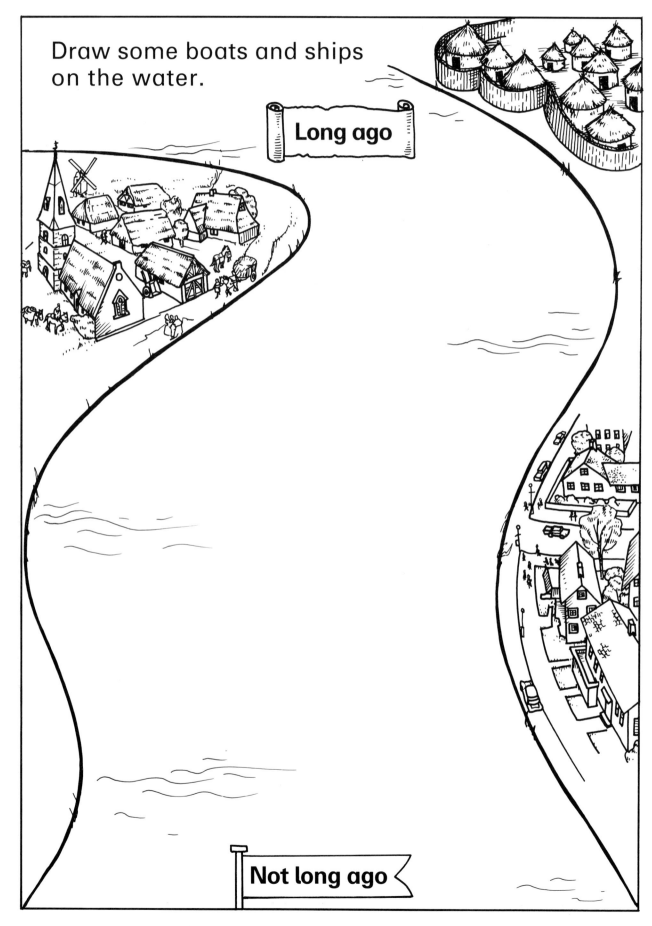

Long ago

Not long ago

Information about jobs

Job	Name of person

Changes in farming

Draw or write about 3 changes in farming.

1

2

3

Craftwork

This is the craft called _____ .

A person who does it is called a _____ .

This is how it is done. (draw or write)

How did this person learn the craft?

What tools are used?

Are there still people doing this craft?

A look at some crafts

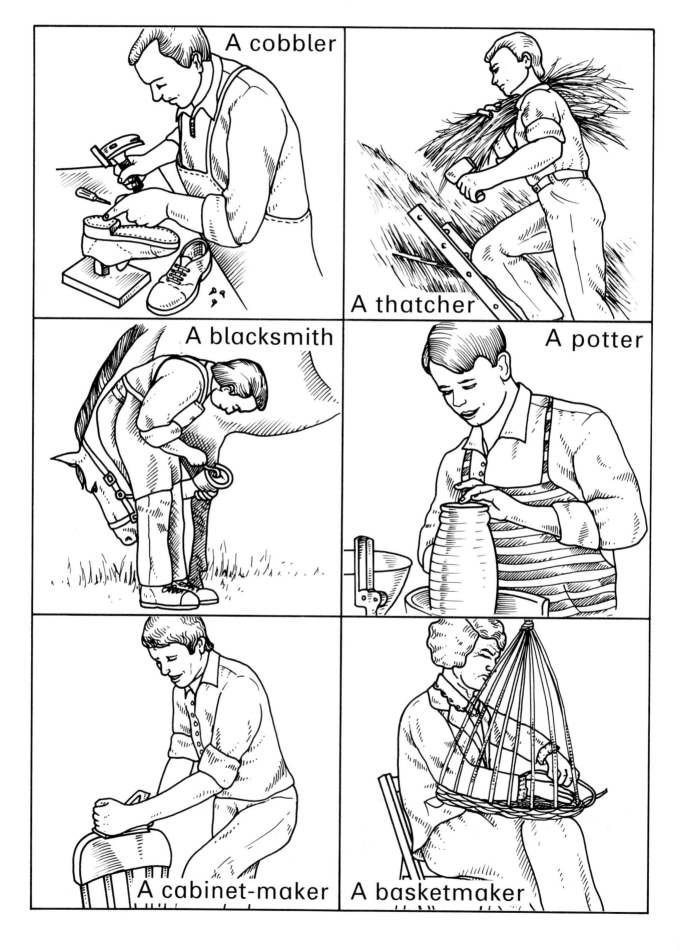

A cobbler

A thatcher

A blacksmith

A potter

A cabinet-maker

A basketmaker

Working as a servant

Why I should not like to be a servant.

Don't run.

Come here.

Do this!
Do that!

You are late again.

Be silent!

Answer that bell.

This is not good enough.

Inside a Victorian house

Pets

A pet called _____
(draw)

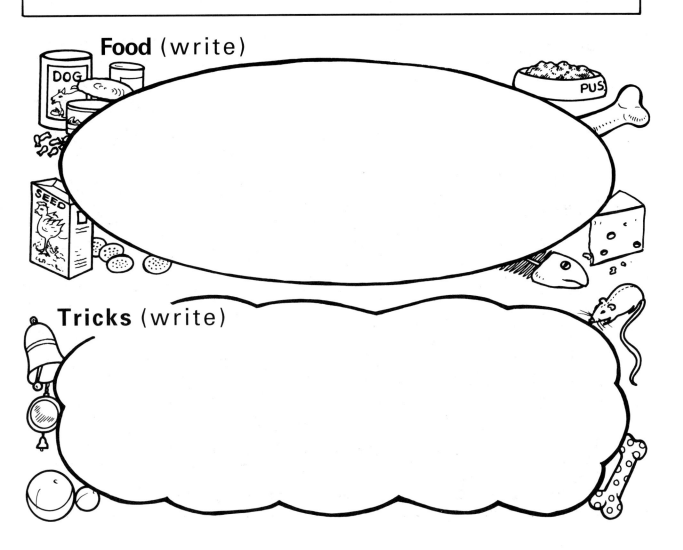

Food (write)

Tricks (write)

Dogs

Pet stories

A pet story I like ...

Facts about pets in the past ...

Horses then and now

Animal designs

Thinking about shopping

The story of the _____ I buy.

Barter

Shopping in the past

Who I talked to _____

When you were young where did your Mum
go shopping?

_____

_____

What sorts of shops were there? _____

_____

_____

Tell me about a shop that is not the same
as it was when you were young.

_____

_____

_____

A medieval market

A day at the market

Draw things on sale.

Draw the stallholder.

Draw what the stallholder wants to take home.

Families and their needs

The Cost of Living in 1950

Essential weekly spending

Food £2   Heat and Light £1   Clothing 50p
Shelter £1·25 rent **OR** £2·75 mortgage

Other essential costs

Weekly pocket money; Mike's bus fares; John's travel to work.

Prices of other items

Travel		Household goods	
Bicycle £12		Cooker £16	Bed £30
Car £575		Fridge £50	Table £20
Petrol 15p (gallon)		Vacuum	Chair £10
Road tax £12		cleaner £30	Lounge
Bus fares $\frac{1}{2}$ p		Fire £10	suite £50
per mile		Washing	Bedroom
		machine £30	suite £90

Entertainment and Leisure

Newspapers 5p a week   Radio £25 (plus £1
Comics 7p a week   licence per year)
Cinema 10p (adult)   TV £75 (plus £3
              5p (child)   licence per year
Bag of sweets about 2p   Record player £14
Tube of fruit gums 1p   A family seaside holiday
                                                 in Britain for 1 week
                                                 approximately £20

The Cotton Family in 1950

John Cotton, aged 35, is a skilled tool-maker. He works 7 miles away in the nearest town and earns £7 a week.

Winnifred Cotton, aged 32, has a part-time job locally. She earns £2.50 a week.

Lucy Cotton, aged 12. She walks to school. Lucy gets 10p a week pocket-money.

Mike Cotton, aged 10. He catches the bus to and from school, which is 4 miles away. Mike also gets 10p a week pocket-money.

The family's essential needs are:

Food   Heat and Light   Clothing
Shelter (they must pay rent or a mortgage)

Other needs (not essential) are:

Entertainment and Leisure
Holidays
Home improvement

From grocery store to supermarket

All about
Harvest Festival

A Victorian Christmas

Choose your own celebration

A special day

C74

I	II	III	IV	V	VI	VII	VIII	IX	X
1	2	3	4	5	6	7	8	9	10

L	C	D	M
50	100	500	1000

Write these in Roman numerals.

Your age _____

12 _____

The number of people in your family _____

Your lucky number _____

The year _____

The number of legs on a chair _____

The number of faces on a cube _____

24 _____   1672 _____

606 _____   73 _____

58 _____   99 _____

A closer look at clocks and watches

Sundials (draw and write)

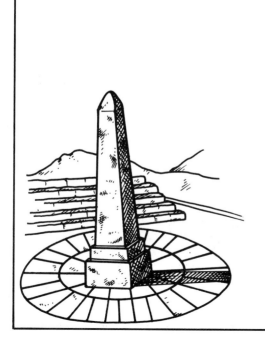

Measures in the past

Measuring using our bodies.

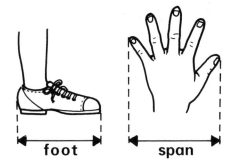

foot   span

stride   reach   arm   cubit

Big clocks

All about _____

(name of clock)

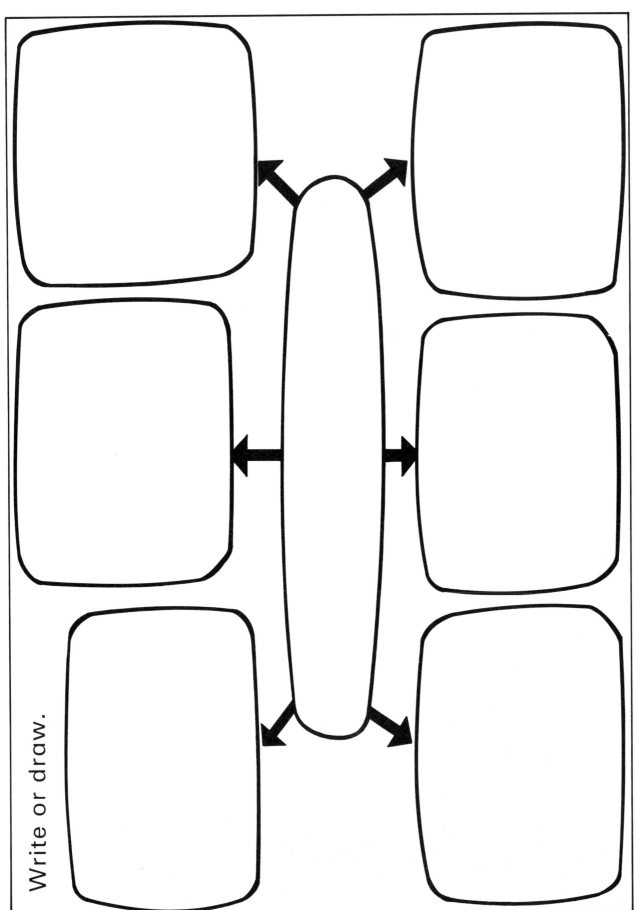

Write or draw.

I looked at _____

This is what I saw.
(write or draw)

Asking questions

Questions I want to ask.

A Victorian street scene

A Tudor street scene

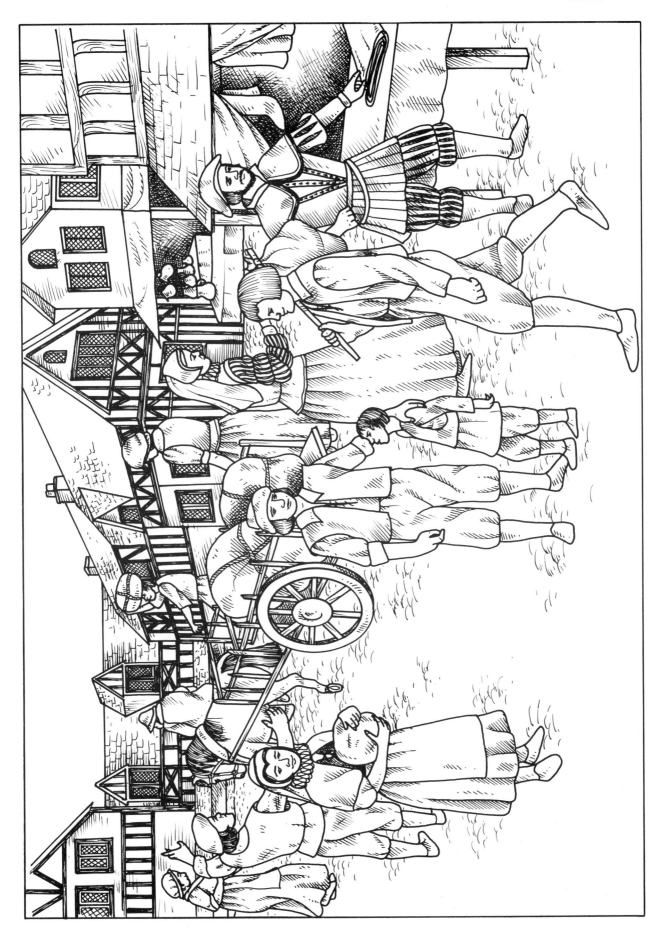

Library skills

A library quiz

How are the books sorted out in your school library?

_____

Story books are sometimes called 'fiction'.
What are information books sometimes called?

_____

Find a history book in the library.
What is its title?

_____

Keep a record of the history books from the school library that you think are useful.

Title	Why it is a useful book

Invaders in Britain

Medieval and Tudor times

Stuart times and Georgian rural life

The Industrial Revolution

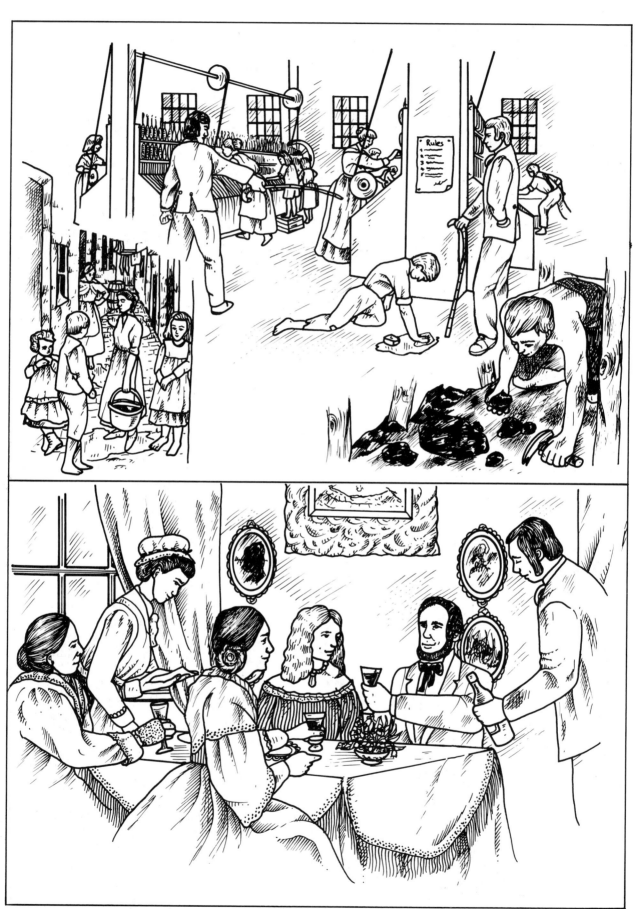

Victorian Britain

After the Second World War

A grown-up I know well _____

_____

Write or draw the five most important things you will tell **your** grandchildren about this grown-up.

What's wrong?

What's wrong?

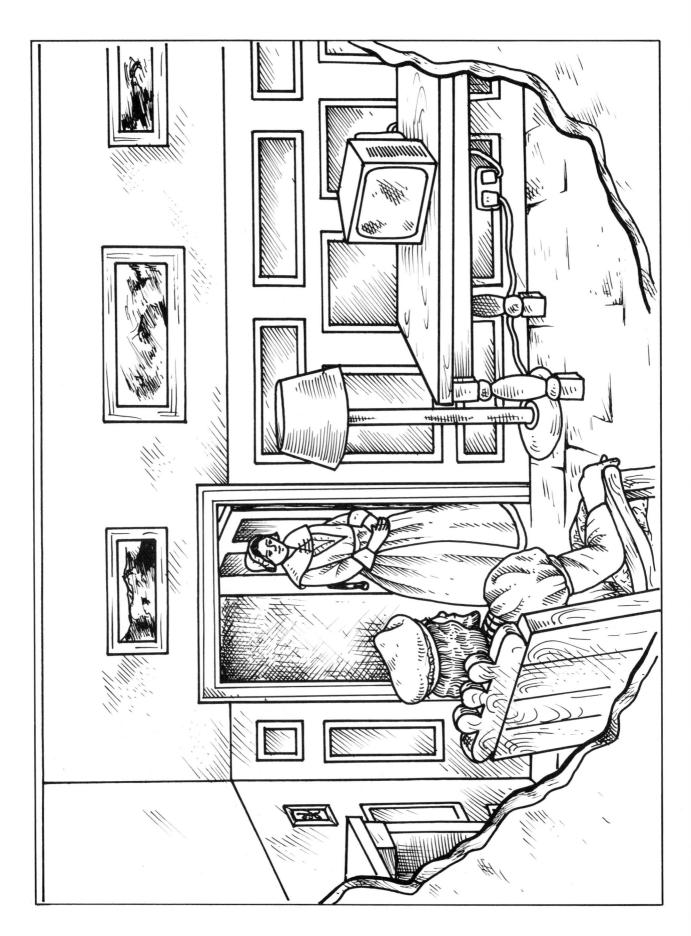

What's wrong?

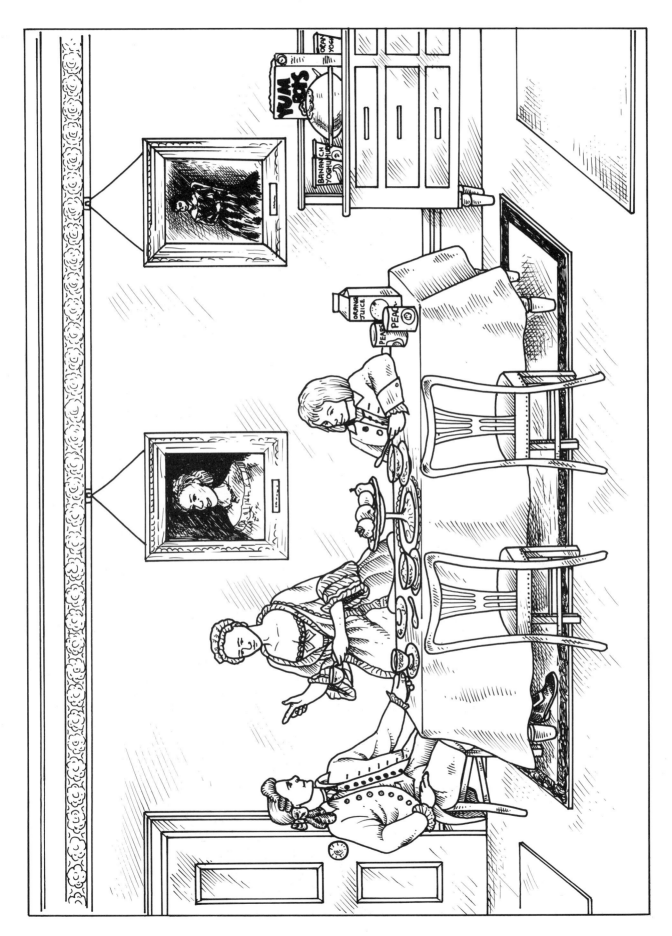

What's wrong?

Self-appraisal sheet

Name _____

When you can do a thing, tick the box. ⟶
Then take this sheet to your teacher.

Teacher's initials

I can set myself a question to answer about the past.		
I can draw an idea web.		
I can observe well.		
I can give an opinion and listen to others.		
I can think of questions to ask about the past.		
I can write down questions to ask about the past.		
I can find clues about the past in pictures.		
I know what the parts of a book are.		
I can find a history book I want in the library.		
I can get information from a database.		
I can put information in a database.		
I can find important clues in pictures and say why I think they are important.		
I can find important clues in stories and say why I think they are important.		
I can talk about differences between two reports about the same thing.		
I can tell other people about my discoveries.		
I can draw things from the past to show what they looked like.		
I can write in a way that is best for what I want to say (diary, story, poem, report).		
I can do research on my own.		

Summary sheet

Name of child _____

Date _____

Topic _____

Level and comments

AT 1	
AT 2	
AT 3	

Date _____

Topic _____

Level and comments

AT 1	
AT 2	
AT 3	

Date _____

Topic _____

Level and comments

AT 1	
AT 2	
AT 3	

Date _____

Topic _____

Level and comments

AT 1	
AT 2	
AT 3	